Keita Ono

XFEM-based Adaptive Contact Model for Telepresence Systems

Keita Ono

XFEM-based Adaptive Contact Model for Telepresence Systems

Time Delay Compensation and Stability Improvement during Medical Incision

Südwestdeutscher Verlag für Hochschulschriften

Impressum/Imprint (nur für Deutschland/only for Germany)
Bibliografische Information der Deutschen Nationalbibliothek: Die Deutsche Nationalbibliothek verzeichnet diese Publikation in der Deutschen Nationalbibliografie; detaillierte bibliografische Daten sind im Internet über http://dnb.d-nb.de abrufbar.
Alle in diesem Buch genannten Marken und Produktnamen unterliegen warenzeichen-, marken- oder patentrechtlichem Schutz bzw. sind Warenzeichen oder eingetragene Warenzeichen der jeweiligen Inhaber. Die Wiedergabe von Marken, Produktnamen, Gebrauchsnamen, Handelsnamen, Warenbezeichnungen u.s.w. in diesem Werk berechtigt auch ohne besondere Kennzeichnung nicht zu der Annahme, dass solche Namen im Sinne der Warenzeichen- und Markenschutzgesetzgebung als frei zu betrachten wären und daher von jedermann benutzt werden dürften.

Coverbild: www.ingimage.com

Verlag: Südwestdeutscher Verlag für Hochschulschriften GmbH & Co. KG
Heinrich-Böcking-Str. 6-8, 66121 Saarbrücken, Deutschland
Telefon +49 681 37 20 271-1, Telefax +49 681 37 20 271-0
Email: info@svh-verlag.de

Approved by: Munich, Technische Universität München, Dissertation, 2011

Herstellung in Deutschland (siehe letzte Seite)
ISBN: 978-3-8381-3364-5

Imprint (only for USA, GB)
Bibliographic information published by the Deutsche Nationalbibliothek: The Deutsche Nationalbibliothek lists this publication in the Deutsche Nationalbibliografie; detailed bibliographic data are available in the Internet at http://dnb.d-nb.de.
Any brand names and product names mentioned in this book are subject to trademark, brand or patent protection and are trademarks or registered trademarks of their respective holders. The use of brand names, product names, common names, trade names, product descriptions etc. even without a particular marking in this works is in no way to be construed to mean that such names may be regarded as unrestricted in respect of trademark and brand protection legislation and could thus be used by anyone.

Cover image: www.ingimage.com

Publisher: Südwestdeutscher Verlag für Hochschulschriften GmbH & Co. KG
Heinrich-Böcking-Str. 6-8, 66121 Saarbrücken, Germany
Phone +49 681 37 20 271-1, Fax +49 681 37 20 271-0
Email: info@svh-verlag.de

Printed in the U.S.A.
Printed in the U.K. by (see last page)
ISBN: 978-3-8381-3364-5

Copyright © 2012 by the author and Südwestdeutscher Verlag für Hochschulschriften GmbH & Co. KG and licensors
All rights reserved. Saarbrücken 2012

Acknowledgment

This dissertation was written when I was working as a scientific assistant at the Institute of Applied Mechanics, Technische Universität München, in Munich Germany as part of the Collaborative Research Centre 453 (SFB453) - *High-Fidelity Telepresence and Teleaction* under subproject M7 supported by the German Research Foundation (DFG). This research work would not have been accomplished without tremendous supports from many persons.

My deepest gratitude belongs to Professor Heinz Ulbrich, *mein Doktorvater* for the opportunity to work under his compassionate supervision. His trust and his teaching had effortlessly enlightened and encouraged me throughout my time at the institute. There is literally no such word which can describe my genuine gratefulness toward him.

I would like to express my sincere thankfulness toward Professor Bodo Heimann, not only for his co-supervising this dissertation, but also for his supports during my time at Leibniz Universität Hannover, in Hannover Germany. Also, I would like to express my gratitude toward Professor Gunther Reinhart for all of valuable discussions during this research project and finally for his duty as the chairman during the oral defense of my dissertation. Moreover, to Professor Friedrich Pfeiffer whom I wish to express my highest gratitude for all of the inspirational and gracious discussions.

To all of my beloved colleagues and personal at the Institute of Applied Mechanics, I wish to thank everyone for their remarkable supports and gorgeous friendship. Especially to Doctor Thomas Thümmel for his fabulous guidance. For the colleagues who spent extensive amount of time to preview and critique the manuscript of this research work; Gerhard Schillhuber, Alexander Ewald, Thomas Villgrattner, Jörg Baur and Johannes Rutzmoser. Also to all students involved in project M7, this research work would not have been finished without all of your effortless contributions.

Finally, I would like to thanks my family and my love for their infinite supports and faith in me. Particularly my mother and my father who have been loving and encouraging me to pursue every dream I ever had.

Keita Ono
June 2012
Ogaki, Japan

Abstract

An incision is a common process in medical telepresence applications. When a large distance between the human operator and the teleoperator is presented, it can lead to a time delay in communication channel which causes the hand's movement and the force feedback perception not to synchronize. This research work proposed an adaptive contact model based on the Extended Finite Element Method. The proposed contact model compensates the time delay using the real-time dynamic geometry deformation simulation and the calculation of the corresponding incision force between the scalpel at the end-effector of the teleoperator and the remote environment. An adaptive parameter identification algorithm is also developed allowing online model verification during the actual incision. The experimental results demonstrate a stability improvement during the incision with the experimental telepresence system.

Keywords: Telepresence system, Extended Finite Element Method (XFEM), Realtime Incision simulation, Adaptive control, Mechatronics

Kurzfassung

In medizinischen Telepräsenzanwendungen können bei großer Entfernung zwischen Operator und Teleoperator Zeitverzögerungen in der Kommunikation dazu führen, dass die Handbewegungen des Operators und die Kraftrückkopplung vom Kraftsensor nicht synchron sind. Zur Lösung des Problems wird ein Kontaktmodell entworfen, das auf der Extended Finite Element Methode basiert. Mit diesem Kontaktmodells ist es möglich die dynamische Verformung eines weichen Kontaktes zu simulieren und die Kraftrückkopplung während des Schnittes in Echtzeit zu berechnen und auszugeben. Dadurch bleibt die Synchronisation zwischen der Operatorbewegung und der Kontaktkraftwahrnehmung erhalten. Zusätzlich wird ein adaptiver Parameteridentifikationsalgorithmus entwickelt und mit dem Kontaktmodell gekoppelt. Damit wird eine Modellverifizierung in Echtzeit ermöglicht und gegebenenfalls eine Adaption der Modellparametereinstellung während der Teleoperation durchgeführt.

Stichworte: Telepräsenzsystem, Extended Finite Elemente Methode (XFEM), Echtzeitsimulation eines Schneidevorgangs, Adaptivregelung, Mechatronik

Contents

1 Introduction   **1**
    1.1 Problem statements and research focuses 2
    1.2 Aspects and outline 3

2 Contact model for time delay in a medical telepresence system   **5**
    2.1 Medical telepresence system 5
    2.2 Telepresence system and time delay 6
    2.3 Contact model and model based control 7
       2.3.1 Smith predictor 7
       2.3.2 Time delay compensation 8
       2.3.3 Modeling of incision in soft body 9
       2.3.4 Family of finite element methods 9

3 Incision mechanics in a soft body   **12**
    3.1 Definition of a cut in a solid body 12
    3.2 Kinematics 13
       3.2.1 Deformation 13
       3.2.2 Material structural and spatial configurational changes 15
       3.2.3 Deformation gradient 15
       3.2.4 Stress and strain 16
       3.2.5 Constitutive equations 17
    3.3 Conservation of momentum 18
    3.4 Boundary value problems 19
       3.4.1 Initial value problems 21
    3.5 Remapping of the governing system of equations 21
    3.6 Weak form of conservation of momentum 22
       3.6.1 Strong to weak form 23
       3.6.2 Smoothness of the test function and the trial function 24
       3.6.3 Principle of virtual work 24

4 Finite element approach   **26**
    4.1 Spatial discretization 26
    4.2 Discrete governing system of equations 28
    4.3 Element coordinates 30
    4.4 Assembly and remeshing 32
       4.4.1 Assembly process 32
       4.4.2 Remeshing 33
    4.5 Dynamic simulation 34
       4.5.1 Newmark implicit integration 34

	4.5.2 Solution of linear algebraic equations	36
4.6	Implementation	37

5 Extended finite element approach 41
5.1 Displacement approximation with discontinuities 42
5.2 Modeling of strong discontinuous fields 43
    5.2.1 Heaviside function 44
    5.2.2 Shifted function 44
5.3 Discrete governing system of equations with XFEM 47
5.4 Enrichment nodes selection and assembly 51
    5.4.1 Enrichment nodes selection 51
    5.4.2 Assembly of enrichment DoFs 52
5.5 Dynamic simulation 53
5.6 Implementation 53

6 Remarks on implementation practicality 57
6.1 Finite element method with remeshing 57
6.2 Extended finite element method 58
6.3 Discussion on the practicality 59

7 Adaptive empirical incision force model 61
7.1 Force modeling of haptic incision perception 61
    7.1.1 Friction on lubricated surface 62
    7.1.2 Incision force model in Cartesian system 63
7.2 Adaptive parameter identification 66
7.3 Optimization 72
7.4 Implementation of the adaptive contact model 73

8 Experiments and results 76
8.1 Experimental telepresence system 76
    8.1.1 Teleoperator 78
    8.1.2 Tension platform 79
    8.1.3 Test object 80
    8.1.4 Software framework 81
8.2 Time delay compensation utilizing contact model 82
    8.2.1 Impact of time delay during incision 83
    8.2.2 Incision force compensation utilizing contact model 86

9 Conclusion and future works 90
9.1 Discussion 90
9.2 Outlooks 91

A Mechanical properties of silicone 93

Bibliography 94

to my parents

1 Introduction

The advancements in engineering and information technologies keep opening new possibilities to the robotic researchers. Therefore, the robotic technology has been evolving in recent years to cover numerous of applications. As one of an important robotic research field, the telerobotic benefits directly from the modern communication technologies allowing a robotic system to be controlled from another location. The telepresence system belongs to the telerobotic category but differentiate itself by a distinguish and control method.

The concept of telepresence and often including teleaction [48] comprises with the essential components which are a human operator, a control devices, a teleoperator and a remote environment which is sensed and acted upon by the teleoperator [41]. The control devices of a telepresence system can be called a human-robot control interface [34]. A simplified schematic outline of a telepresence system is presented in Figure 1.1.

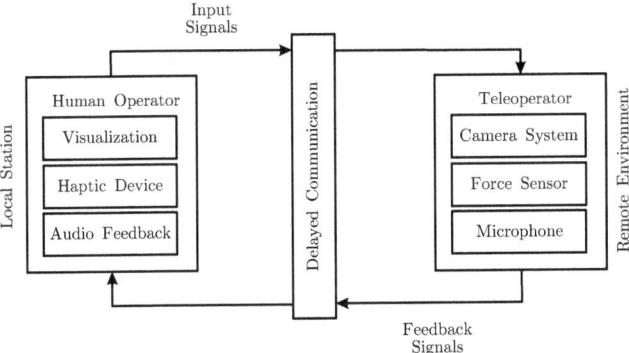

Figure 1.1: A schematic of a multimodal telepresence system, adapted from [65]

A telepresense system employs a bilateral control scheme in order to provide a human operator a virtual but realistic access to a remote environment at distance. The devices used as the control interface, for instances, a haptic device, a head-mounted display and a three-dimensional audio feedback are generally mapped to the corresponding sensor arrays of the teleoperator in the remote environment. The advantage of a telepresence system is that the existing of the human operator in the same location of the robotic system may not be necessary. The control interface can be intuitive and connected to the sensorial system of a human operator. Which means that this human operator is associated with the teleoperator system in the remote environment via a network infrastructure and interactively manipulates the remote environment with dexterity relying on the feedback signals. Because of its control interface, the telepresence allows theoretically a human operator to transfer their particular skill to the robotic system. It can be alleged that if the telepresence experience is natural or transparent to a human operator completely, the rate of success in the given task shall solely depend on their skill factors. The ideal telepresence is often described as an immersive experience of the human operator into the remote environment with all his or her sensory abilities.

Telepresence systems are involving in different areas of applications. For a short example: the space exploration and on-orbit service, the ROKVISS of the German Aerospace Center [50, 69] and the Robonaunt of NASA [47], medical treatments [44, 53, 64], plant maintenance and assembly [22, 70]. The researchers in this area have not been being challenged only to overcome technical limitation of the system implementation but also forced to encounter a requirement of producing a reliable robotic system.

As an example, Collaborative Research Centre 453 (SFB 453) *High-Fidelity Telepresence and Teleaction* (1998-2010), financially supported by the *German Research Foundation* (DFG) was appointed particularly to explore the possibilities and develop new technologies of the telepresence applications in numerous branches of areas [1].

1.1 Problem statements and research focuses

An important characteristic of a telepresence system is the distance between the teleoperator and its human operator. This advantage is a self-contradictory since it may lead to a time delay in the communication channel. A time delay exists in each communication medium and, if significant, is one of major causes to the failure of telepresence system causing the loss of the synchronization between the input commands and the feedback perceptions. To solve this technical problem, this research work, under subproject M7 of SFB 453, proposed the principle of contact model to compensate the time delay in the communication. A contact model is required to substitute the actual delayed feedback visualization and the contact force perception with the results from a real-time simulation.

The contact model developed in this work focuses specifically on the incision process in a soft body with a scalpel common for the medical telepresence applications. To achieve a real-time graphic visualization simulation, a dynamic geometry deformation of the soft body must be demonstrated. Two approaches, Finite Element Method (FEM) and Extended Finite Element Method (XFEM) are investigated. The calculation of the contact force between a scalpel and the soft body is done using the empirical incision force model.

Additionally, the usability of the contact model is enhanced by the introduction of an adaptive parameter identification algorithm. The algorithm allows the contact model to be verified online during the actual incision is being carried out by the teleoperator.

1.2 Aspects and outline

To achieve this research focuses, the outline of this work is systematically designed to open essential discussions related to the topic while providing an answer to each discussion from the development of the proposed adaptive contact model to the conclusion of the results.

To begin with, Chapter 2 is developed to provide an overview of the primary problem for a medical telepresence system regarding time delay. The essential engineering concepts and principles employed in the development will be presented along with the introductory notes on important technical definitions with their related works commonly referred in this specific area of applications.

Chapter 3 will extensively discuss on the physical interpretation of the incision mechanics using the theory of continuum mechanics. The material and structural changes due to the evolution of the incision will be considered. The investigation in this chapter provides the linear governing system of equations necessary for the dynamic simulation of the geometry deformation of a soft body subjected to incision.

To achieve a real-time simulation of the geometry deformation, the derived linear governing system of equations is discretized using FEM in Chapter 4. A discussion on the remeshing algorithm required to update the governing system of equations corresponding to the cut evolution will be made. Chapter 5 is devoted to XFEM which is the second approach employed to discrete the governing system of equations derived in Chapter 3. The demonstration of a cut evolution by XFEM in term of a mathematic approximation will be presented and discussed.

FEM and XFEM approaches are investigated in this work as a implementation tool for a contact model for incision process. It is critical that the approach should be practical and flexible for implementation a different physical aspects of the unfamiliar remote environment. In Chapter 6, the evaluation on this matter will be specifically made.

The development of an empirical incision force model is demonstrated in Chapter 7 along with its adaptive parameter identification algorithm. Additionally, the optimization employed to the performance of the incision force model the will be presented.

In chapter 8, the results of the impact of time delay to the test object during an incision will be presented in comparison with the results of the same incision process

1.2 Aspects and outline

while utilizing the proposed adaptive contact model. Chapter 9 concludes the results with related discussions and outlooks.

2 Contact model for time delay in a medical telepresence system

This chapter is aimed to provide an overview of topics related to the problem statements and the focuses of this work. The chapter will begin with a brief introduction of the medical telepresence system before concentrates on a problem of time delay in the communication. The principle and benefits of contact model will be discussed in terms of classical control theory as a feasible solution to time delay problem. The theory of continuum mechanics and Finite Element Methods employed in this work as development tools of the contact model for an incision process will be also discussed.

2.1 Medical telepresence system

There are many telepresence systems developed not only allowing skill transfer from a human operator to the machine but also to enhance and improve the performance beyond their ordinary abilities. Particularly in medical area, a robot based surgery is one of major interests of many telepresence developers. A medical telepresence system is generally designed so that the surgeon can handle the same operation using their individual skills and techniques. The ultimate goal in medical telerobotic is to allow a surgeon to provide his medical expertises to a patient independent from location.

It is however a fact that the current generation of telepresence systems do not yet provide an ideal and transparent telepresence experiences to their operators. This is generally because the different in physical of the robot manipulators and human arms with the control interface. Therefore, a direct skill transfer from a human to machine can not be easily achieved. The researchers take an effort to develop the supplemental systems helping the operators to overcome the limit or unfamiliarity of the control interface. For example, TAVAKOLI ET AL. investigates the different in joint impedances between the control interface and the teleoperator manipulator of their Zeus medical robot system. They proposed a sensor which measures the velocity of the robot end-effector and accordingly calculates a compensation force at the operator's side which matches their impedances [90]. HIRZINGER ET AL. develops MiroSurge aiming for heart surgery capable of automatic manipulator compensation for heart-beating movement [44].

The control interface of the telepresence system may not be intuitive enough in some cases for its human operator to undergone a complicate task-handling. Thus, the automatic routine task-handling and the visual guidance control algorithms is intensively studied by MAYER and STAUB to support and ease the learning-curve of a human operator. The algorithms provides aid to the operator with an semi-automatic sequence by knot-typing [53, 87].

Some of development criteria are not focusing on the performance enhancement alone, but also pursuing to maintain or improve the usability and efficiency of the telepresence system. For instance, time delay in communication channel can reduce the level of fidelity of the telepresence system and prevent the operation to be carried out successfully. The main focus of this work stays in this particular category.

2.2 Telepresence system and time delay

In some telepresence systems, a technical limit such as large time delay in communication is inevitable. Significant time delay occurs if the distance between the human operator and the teleoperator is large or a communication protocol with inefficient transmission rate is involved. The time delay in a telepresence system causes a postpone arrival of the input and feedback signals. This causes the level of synchronization between these signals to be declined. Because the telepresence system is controlled by human operators and relying on their perceptions, the human operators may react to the false feedback perception. In a severe case, the remote environment and the teleoperator can be damaged. It is difficult to justify the maximum amount of time delay which still allows the telepresence operation to be safely executed. It depends strongly on the individual application, hardware design and foremost, human factors. As a guide line quoted by HELD and SHERIDAN, $300\,ms$ is often referred as a maximum allowance [41, 76].

Most of the current telepresence systems require an individual communication channel, e.g. over satellite or fiber optic cable. However, the approach only promises the communication will not be interrupted easily by an unforeseen transmissions. Thus, the data transfer rate should be stable as far as the physical limits of the network can provide. On the other hand, ECKEHARD proposes one of a logical step to increase the stability by reducing the amount of transmitted data [88]. However, in random situations where the data are completely lost due to an interruption in communication, the reduction of data may cause a severe damage since the reduced or compressed data by the algorithm may contain valuable information. The principle of wave variable is one another well-known method employed in many current telepresence systems to especially cope with the time delay problem. Example implementations on this topic can be referred by the works of SLOTINE ET AL. [82, 83], HIRSCHE ET AL. [42, 43] also SMITH ET AL. [84]. This principle alters both input and output variables transmitted fort and back between operator and the teleoperator to guarantee a certain level of synchronization between the feedback perception and the actual movement of the teleoperator. However, its effect may conduct an unrealistic telepresence experience due to the incorrect rendering of the actual contact behavior. For this reason, the principle of wave variable may not suitable for a telepresence scenario such as medical teleoperation involving incision process.

The second fact about time delay demonstrated by MITSUISHI is its dependency on the physical distance between the operator and the teleoperator comprising the networking and individual equipments used [55]. It is also true that networking required many levels of data synchronizations and package loss are common. Shorter physical distance and less number of medium equipments are definitely preferred. In most case, time delay is treated as a property of the system. It should be known at hand before initiating a telepresence system. Consider the impracticality of reserving a satellite communication channel or adapt the communication protocol including its networking equipments, the ideal communication sufficient for a transparent telepresence experience to the operator can be questionable.

2.3 Contact model and model based control

2.3.1 Smith predictor

In a classical control, the time delay problem has been addressed for decades. A large time delay in a control loop is seen frequently in many process control applications, such as a chemical process in which a real-time monitoring of actual chemical reaction is often not feasible. With a time delay existing between the plant and the controller, the gain-margin and phase-margin are decreased. It is up to the plant dynamic when should the system goes instable. The actual causes depend on the process itself and partly due to sensor limitations. In other words, the monitoring of the state of the process can not be measured physically due to a sub-reaction with a large time-constant prohibits a proper measurement of the actual state of the process before the control substance reaches the sensitivity range of the sensor. larger the time delay is, will the system more likely be unstable. This phenomenon is comparable, in case of a telepresence system, to the delay of data arrival and data lost in its communication channel, since a prompt arrival of the feedback variables are not exactly guaranteed.

The Smith predictor, depicted in Figure 2.1 genuinely introduced by O.J.M. SMITH in 1957. There are different implementations and the exact control scheme of Smith predictor is not conclusive because of differences in individual control requirements. However, it is absolute that the method proposes a solution to time delay problem by allowing the controller to respond to the simulated feedback signal instead of solely to the actual feedback signal. A simulated or predicted feedback signal is obtained by the model of the actual process but without physical time delay. An actual feedback signal can be generally required for the controller to response if the residual error in the control loop remains. With this principle, the stability of the overall control loop can be improved accordingly due to reduction in task completion time [8,84,85], provided that the contact mechanics or the actual process is known.

2.3 Contact model and model based control

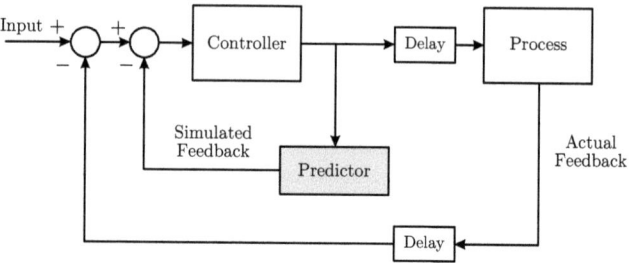

Figure 2.1: A control scheme of Smith predictor

Theoretically, a predictor can achieve a superior performance to feedback-based control techniques due to minimum compromising and intervention to the input variable. It also provide a flexibility in the development of a control scheme since the actual delayed feedback is not always necessary.

2.3.2 Time delay compensation

By adopting concept of Smith predictor in order to compensate the time delay from the feedback loop, a contact model is integrated into the conventional control scheme of a telepresence system as a predictor as demonstrated in Figure 2.2. In this context, a contact model is a real-time simulation of the dynamic contact behavior between the end-effector of the teleoperator and its remote environment. It provides the vital feedback signals which are a predictive visualization and the incision feedback to its human operator via monitor and haptic device. The contact model is appointed to permanently provide correct feedback variables in order to eliminate the actual delayed feedback signals completely. The modeling of geometry deformation is done relying on the theory of continuum mechanics and Finite Element Methods. Whereas, the incision force is calculated from the empirical incision force model. This work also proposed an adaptive parameter identification which instead uses the delayed force feedback to verify the correctness of the empirical force model. If the error in the empirical force model occurs its model parameter can be adjusted during the actual incision.

As of telepresence system, Smith predictor takes benefit from precise and efficient modeling techniques in order to simulate the feedback signal in real-time, hence the system stability is maintained. Depending on the type of teleoperation, a contact model can be modeled using different principles. GOLLE implemented a contact force prediction using multi-body with non-smooth contact theory for teleoperation with rigid bodies [32, 33]. SCHILLHUBER developed the contact force prediction algorithm for a contact between the end-effector and a deformable body based on

2.3 Contact model and model based control

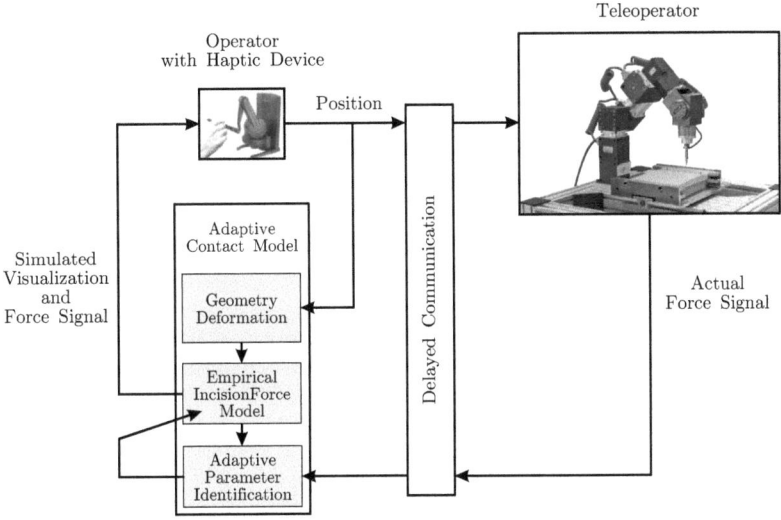

Figure 2.2: Telepresence control scheme with adaptive incision force compensation algorithm

non-linear finite element method [71–73]. ZHAO also demonstrates an implementation of contact model with inhomogeneous materials [100].

2.3.3 Modeling of incision in soft body

The development of contact model in this work concentrates specifically on an incision process in a soft body. Doing so, the analytic and mathematical interpretation of the incision mechanics and the interaction of the cutting tool and the soft body must be found. Modeling of contact mechanics has its root from continuum mechanics and multi-body simulation. For studying and understanding purpose, the works contributed by PFEIFFER [66], ULBRICH [95] and CLOUGH [23] exhibits lectures and engineering examples of the linear and non-linear formulation in practical mechanic problems.

2.3.4 Family of finite element methods

In many situations, an adequate model derived from its classical theories is obtained using a finite number of well-defined components. Such problems are also known as discrete. Discretization of a continuous problem keeps happening because of

2.3 Contact model and model based control

the presentation limitations of a complex behavior in one operation [102]. From classical theories of mechanics to the actual implementation, the family of finite element methods keep repeating their success as an essential simulation of tool in this very computing age. The development of the finite element has become a principle with scientific contributions in every way. As for engineers, the discretization of continuous problems has been approached intuitively by creating an analogy between real discrete elements and finite portions of a continuum domains. This statement is also true to the approaches discussed in this research work.

In the early time of developing the principle of finite element especially in the field of solid mechanics, MCHENRY [54] HRENIKOFF [45] NEWMARK [60, 61] and SOUTH-WELL [86] showed evidences that reasonably good solutions to an elastic continuum problem can be obtained by replacing small portions of the continuum by an arrangement of simple elastic bars. Later, TUNER ET AL [94]. demonstrated a more direct substitution of properties can be made effectively by considering small portions or elements in a continuum behaving in a simplified manner. One of the most important contributions was given by ZIENKIEWICZ AND TAYLOR [101, 102] who gathered the necessary and relevant mathematical analysis into what would be called the Finite Element Method (FEM), building the pioneering mathematical formalism of the method. The books of BATHE [13], Wriggers [98, 99], BELYTSCHKO [15] and EBERHARD [26] are suitable for introductory but crucial examples to engineering finite element applications and problem formulations.

The first effort for developing the Extended Finite Element Method (XFEM) methodology can be traced back to 1999 when BELYTSCHKO AND BLACK presented a minimal remeshing Finite Element Method for crack growth [14]. They added discontinuous enrichment functions to the Finite Element approximation to account for the presence of the crack. This method allows the crack to be arbitrarily aligned within the mesh. MOËS later improved the method and called it the Extended Finite Element Method [56, 57]. A major contribution in this area can also be found in the dissertation of DOLBOW [25]. XFEM has been adopted into number of problems involving the discontinuities in the element while requires minimum consideration on the mesh condition [29]. The minimum consideration of mesh condition is known as the main advantage of XFEM in contrast to FEM which generally requires reconsideration of mesh when treated with discontinuity. The more recent and profound works in this field can also be referred to ones of SUKUMAR, BORDAS, HANSBO AND HANSBO and FRIES [16, 29, 39, 89]. MOHAMMADI has published an introduction to XFEM comprising a range of enrichment methods for different discontinuity types in solid mechanics [58].

Both FEM and XFEM are popular as implementation tools of real-time interactive geometry deformation simulations. Since the simulation of the deformation of such complex bodies requires a discretization to be efficient. Both methods are also suitable to the computing and graphic visualization which demonstrates a full body as a finite number of elements or triangles. TERZOPOULOS exhibited a dynamic simulation of a soft-body and the rendering of the contact force allowing interaction with a haptic device [91]. JEŘÁBKOVÁ demonstrates a use of XFEM and parallelized

2.3 Contact model and model based control

computing on a multi-cored CPU [46] to achieve a simulation of virtual cutting. CONTI ET AL. invents an open-sourced haptic development library CHAI3D which provides a rapid haptic scene developing framework. Similarly, SOFA framework is developed by ALLARD ET AL. combine FEM and XFEM functionality and intuitive haptic interface. A possibility of an automatic performance evaluation of the surgical simulator was demonstrated by SEWELL [75].

The computational expense of the simulation is another challenge preventing a successful realization of contact model or Smith predictor concept in a telepresence system. It is arguable that a contact model should not involve in an complicate contact due to its inefficiency to substitute the feedback variables with a task completion time smaller than the actual time delay in the communication. This statement has been true for a computational point of view. Nonetheless, a simulation of incision especially with a soft-body requires that level of complexity. However, with modern computational techniques and hardwares, a real-time simulation of contact phenomena with multi-level of optimization is in their progress and has been improved dramatically in the recent years. One major aspect which minimize the computational time of such complex simulator is the modern computing architecture, both software and hardware such as parallel computing on multi-cored CPU and graphic card [2,3,5].

3 Incision mechanics in a soft body

A first step in considering a physical model should be a clear and precise elaboration of the purpose of the kinematic model. The different approaches may lead to different results. In any case the chances of establishing a proper model depend strongly on a thorough understanding of the physical processes of the object to be modeled. A good model can be referred as a good representation of mechanical properties and therefore a good correspondence to practice and its measurement [66].

This chapter is the beginning of the main objective of this work, which is the development of a contact model for a medical incision process. For the visualization of the dynamic geometry deformation required by the contact model, a physical model of incision mechanics must be investigated before the governing system of equations can be derived. Doing so, the definition of an incision or a cut in a solid body on the basis of continuum mechanics will be discussed in this chapter. Since the simulation of an incision in a large a solid body is the intention of this work, the derivation of kinematic variables and the governing system of equations using a linear strain measure is sufficient.

Some formulations found in this chapter are also known in fracture mechanics such as derivations of BELYTSCHKO, BLACK, MOËS and GÜRSES [14, 36, 56, 57]. Although they were genuinely developed to handle the fracture and crack growth simulation, their derivations treat cracks and cuts similarly as a strong material discontinuity in a solid body with also applicable for modeling of a cut in a solid body.

3.1 Definition of a cut in a solid body

Incision process can be seen generally as an interaction between a cutting equipment and a solid body to produce a mark or open a surface of a solid body. A solid body $\mathcal{B}$ stays in a fixed reference frame in the Eulerian space $\mathbb{R}^3$ as depicted in Figure 3.1. The cut is defined by a family of time-dependent volume $\Gamma(t) \subset \mathcal{B}$, as

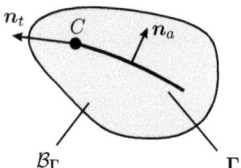

Figure 3.1: Definition of a solid body with a cut (adapted from GÜRSES [36])

$$\Gamma(t_1) \subset \Gamma(t_2) \quad \text{for} \quad t_1 \leq t_2 \tag{3.1}$$

with a smooth boundary $\partial \Gamma(t)$. In crack propagation, this boundary is often referred as the crack tip or front [36]. In a similar manner the definition of a cut says that the cut propagation is allowed only through the motion of the front $\partial \Gamma$. The current cut volume is therefore derived as the set

$$\Gamma(t) = \Gamma(0) \cup \left\{ \partial \Gamma(\tilde{t}) \mid 0 < \tilde{t} < t \right\} \tag{3.2}$$

where $\Gamma(0)$ is the initial cut volume. In the current configuration at time t, particles in body $\mathcal{B}$ occupy the region

$$\mathcal{B}_\Gamma(t) := \mathcal{B} \setminus \{\Gamma(t) \cup \partial \Gamma(t)\} \subset \mathbb{R}^3 \tag{3.3}$$

which have the outer boundary $\partial \mathcal{B}$ and the inner boundary formed by cut. The tangential vector $\boldsymbol{n}_t$ defies the direction of the cut Γ and points where the cut will evolve. The normal vector $\boldsymbol{n}_a$, on the other hand, is perpendicular to the cut volume $\partial \Gamma$. The rate of cut evolve shall be zero on the volume of the cut which is not the cut front C.

3.2 Kinematics

The structural changes and deformation occurs to the solid body $\mathcal{B}$ is depicted by Figure 3.2. The cut in general causes the structural changes in a solid body, which is denoted by Γ on the domain Ω_Γ. The same cut is then mapped as the geometry changes to the solid body domain in reference coordinate known as $\mathcal{B}_\Gamma$. The geometry changes due to cut including *Dirichlet* and *Neumann* boundary condition cause the deformation in the current configuration $\mathcal{S}_\Gamma$.

The coordinate vector $\boldsymbol{X} \in \mathcal{B}_\Gamma \cup \Gamma(t)$ is referred to the material or *Lagrangian* coordinates of a particle in a solid body $\mathcal{B}$ in its reference configuration. In a deformed or current configuration at time t, the same particle is denoted by the vector in spatial or *Eulerian* coordinate $\boldsymbol{x}(\boldsymbol{X},t) \in \mathcal{S}_\Gamma$, where $\mathcal{S}_\Gamma \subset \mathbb{R}^3$.

3.2.1 Deformation

The deformation φ_t can be mapped accordingly by

$$\varphi_t : \begin{cases} \mathcal{B}_\Gamma \to \mathcal{S}_\Gamma \\ \boldsymbol{X} \mapsto \boldsymbol{x} = \varphi_t(\boldsymbol{X}) \end{cases} \tag{3.4}$$

The deformation φ_t is therefore *only valid* for the particles in body $\mathcal{B}$ which are not on the cut and cut boundary $\boldsymbol{X} \notin \{\Gamma \cup \partial \Gamma\}$ at their reference configuration. The

3.2 Kinematics

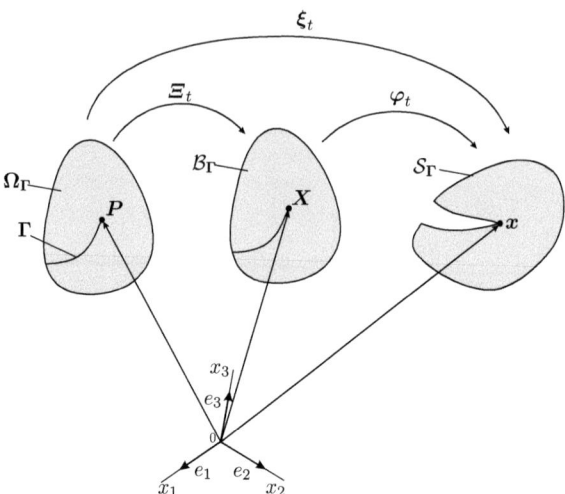

Figure 3.2: Kinematic description of the structural changes and deformation in a body containing a cut

invert of the deformation function exists as

$$\varphi_t^{-1} : \begin{cases} \mathcal{S}_\Gamma \to \mathcal{B}_\Gamma \\ x \mapsto X = \varphi_t^{-1}(x) \end{cases} \quad (3.5)$$

For a particle on the solid body, which in its reference coordinate is pointed by $X \in \mathcal{B}_\Gamma$, the displacement U is known from continuum mechanics as,

$$U(X,t) = \varphi_t(X) - X = x(X,t) - X \quad (3.6)$$

The spatial displacement u in current configuration is defined in the similar fashion,

$$u(x,t) = x - \varphi^{-1}(x,t) = x - X(x,t) \quad (3.7)$$

For both reference and the current configurations staying in the same uniform Eulerian coordinate, their displacement shall be identical.

$$U(X,t) = u(x,t) \quad (3.8)$$

Time derivative of the displacement $U(X,t)$ is known as the velocity field of a particle $V(X,t)$

$$\begin{aligned} V(X,t) &= \dot{U}(X,t) = \dot{\varphi}_t(X) \\ &= \frac{\mathrm{d}(X + U(X,t))}{\mathrm{d}t} = \frac{\mathrm{d}U(X,t)}{\mathrm{d}t} \end{aligned} \quad (3.9)$$

3.2 Kinematics

and for the spatial velocity field $v(X,t)$

$$v(X,t) = \dot{u}(X,t) = V(X,t) \tag{3.10}$$

The second time derivative of the displacement $U(X,t)$ represents the acceleration field of a particle $A(X,t)$

$$\begin{aligned}A(X,t) &= \dot{V}(X,t) = \ddot{U}(X,t) \\ &= \frac{\mathrm{d}V(X,t)}{\mathrm{d}t} = \frac{\mathrm{d}^2 U(X,t)}{\mathrm{d}t^2}\end{aligned} \tag{3.11}$$

and the spatial of the acceleration field $a(X,t)$

$$a(X,t) = \frac{\mathrm{d}v(x,t)}{\mathrm{d}t} = \frac{\partial v(x,t)}{\partial t} + \frac{\partial v(x,t)}{\partial x} \cdot v(X,t)$$

3.2.2 Material structural and spatial configurational changes

A one-to-one piecewise transformation $\Xi_t : \Omega_\Gamma \to \mathcal{B}_\Gamma$ of the reference configuration is introduced to map the material structural changes due to cut in the body $\mathcal{B}$ [36]. This mapping is considered as the time-dependent parameterization of the medium that accounts for the material structural changes such as a cut evolution governed by the cut tip $\partial \Gamma$. It coincides with the time-dependent change from the reference coordinates $P \in \Omega_\Gamma$ to the Lagrangian coordinates $X \in \{\Gamma \cup \partial \Gamma\}$.

$$\Xi_t : \begin{cases} \Omega_\Gamma \to \mathcal{B}_\Gamma \\ P \mapsto X = \Xi_t(P) \end{cases} \tag{3.12}$$

In addition, the spatial configurational change ξ_t is given to map the change occurs in material and structure of the body $\mathcal{B}$ to the actual coordinate of a particle at current time configuration x, as

$$\xi_t : \begin{cases} \Omega_\Gamma \to \mathcal{S}_\Gamma \\ P \mapsto x = \xi_t(P) \end{cases} \tag{3.13}$$

3.2.3 Deformation gradient

In the domain $\mathcal{B}_\Gamma := \mathcal{B} \setminus \{\Gamma(t) \cup \partial \Gamma(t)\}$, where the cut is excluded, the measurement of the local deformation is given by deformation gradient,

$$\begin{aligned}F &:= \nabla \varphi_t(X) = \frac{\partial \varphi_t}{\partial X} \\ &= \nabla x(X,t) = \frac{\partial x}{\partial X}\end{aligned} \tag{3.14}$$

Since F is the derivative of the current position x of a particle at time t with respect to the coordinates of the reference configuration X, it describes the relation between

3.2 Kinematics

a line segment in the reference configuration and the corresponding segment in the current configuration. The deformation gradient can be interpreted as a measure for the deformation of a line segment $\mathrm{d}\boldsymbol{X}$, with

$$\mathrm{d}\boldsymbol{x} = \boldsymbol{F}\mathrm{d}\boldsymbol{X} \tag{3.15}$$

Similarly, there are relations called transport theorems for the deformations of infinitesimal surfaces

$$\mathrm{d}a = J\boldsymbol{F}^{-T}\,\mathrm{d}A \tag{3.16}$$

and volumes

$$\mathrm{d}v = J\,\mathrm{d}V \tag{3.17}$$

The determinant of $\boldsymbol{F}$ recognized from the *Nanson*'s formula

$$J = \det \boldsymbol{F} \tag{3.18}$$

is known as Jacobian and is used to measure the local change of volume.

3.2.4 Stress and strain

The *Cauchy stress vector* $\boldsymbol{T}$ is defined from a traction force vector $\Delta \boldsymbol{f}$ acting on an infinitesimal area ΔA as

$$\boldsymbol{T} = \lim_{\Delta A \to 0} \frac{\Delta \boldsymbol{f}}{\Delta A} \tag{3.19}$$

With the surface normal vector $\boldsymbol{n}$, the Cauchy stress vector is derived as

$$\boldsymbol{T} = \boldsymbol{\sigma} \cdot \boldsymbol{n} = \boldsymbol{\sigma}^T \cdot \boldsymbol{n} \tag{3.20}$$

Tensor $\boldsymbol{\sigma}$ is called *Cauchy stress tensor*. To fulfill the spatial balance of angular momentum, the Cauchy stress tensor is always symmetric and belongs to the current configuration by definition. σ_{ij} where $i = j$ denotes the normal stress, whereas σ_{ij} where $i \neq j$ are shear stress. The stress tensor describes the state of stress in a solid body, hence it must be determined in terms of one of a configuration or coordinate system.

The deformation gradient $\boldsymbol{F}$ is not a suitable measure of the actual deformation of a material point due to its dependence on the motion of a solid body $\left(\dfrac{\partial \boldsymbol{x}}{\partial \boldsymbol{X}}\right)$. The *Green-Lagrange strain* is introduced instead to eliminate this problem. The Green-Lagrange strain tensor is non-linear and often considered as a better measure compared to the deformation gradient since it measures the strain by compare the

3.2 Kinematics

rate of displacement U in the reference configuration X as

$$E^X = \frac{1}{2}\left(\nabla_X u^T + \nabla_X u + \nabla_X u^T \cdot \nabla_X u\right) \quad (3.21)$$

or index-wise

$$E^X_{ij} = \frac{1}{2}\left(\frac{\partial u_i}{\partial X_j} + \frac{\partial u_j}{\partial X_i} + \frac{\partial u_k}{\partial X_i}\frac{\partial u_k}{\partial X_j}\right) \quad (3.22)$$

Obviously, the Green-Lagrange strain tensor is configuration dependent, therefore $E^X \neq E^x$. In context of this work, which is focusing on the modeling of incision process, the deformation is considered to be small with respect to the total geometry of the test scenery. Thus for a small displacement in a solid-body $\|\Delta U\|$, there exists from linearization that a kinematic variable which is partial differentiated with respect to X is

$$\frac{\partial \bullet}{\partial X} = \frac{\partial \bullet}{\partial x} \cdot \frac{\partial x}{\partial X} = \frac{\partial \bullet}{\partial x} \cdot \frac{\partial (X+U)}{\partial X} \approx \frac{\partial \bullet}{\partial x} \quad (3.23)$$

With this linearization, the Green-Lagrange strain tensor is simplified into the symmetric infinitesimal strain tensor ϵ.

$$E^X_{ij} \approx E^x_{ij} = \frac{1}{2}\left(\frac{\partial u_i}{\partial x_j} + \frac{\partial u_j}{\partial x_i}\right) := \epsilon_{ij} \quad (3.24)$$

3.2.5 Constitutive equations

Previously, it was shown that the stresses in a solid body made of elastic material is a result from the deformation of the material. The deformation is also possible to express them in terms of strain. To obtain it, the relation between the stress and strain known as *Hooke*'s material law is brought in Equation 3.25.

$$\sigma = C : \epsilon \quad \text{or} \quad \sigma_{ij} = C_{ijkl}\epsilon_{kl} \quad (3.25)$$

It is capable of providing a explanation of the linear elastic material behavior. The 4th-order strain tensor $C := C^4$ contains $3^4 = 81$ independent elements C_{ijkl}. However, because of the symmetry of both the stress tensor σ and the strain tensor ϵ, the number of independent elements is reduced to 36. Together with the consideration of the strain energy of a linear elastic anisotropic material, the number is again reduced to 21. For an orthotropic material, there are only nine independent elements to be considered. If the material is isotropic, the strain tensor C can be completely determined by two independent elements, the *Young's modulus* E and *Poisson* ration ν, or *Lamé* parameters λ and μ

$$\sigma = \lambda \text{tr}(\epsilon)I + 2\mu\epsilon \quad \text{or} \quad \sigma_{ij} = \lambda\epsilon_{kk}\delta_{ij} + 2\mu\epsilon_{ij} \quad (3.26)$$

with

$$\lambda = \frac{\nu E}{(1+\nu)(1-2\nu)} \quad \text{and} \quad \mu = \frac{E}{2(1+\nu)} \tag{3.27}$$

With equation 3.26, the stress tensor is recognized as

$$\begin{bmatrix} \sigma_{11} & \sigma_{12} & \sigma_{13} \\ \sigma_{21} & \sigma_{22} & \sigma_{23} \\ \sigma_{31} & \sigma_{32} & \sigma_{33} \end{bmatrix} = \begin{bmatrix} 2\mu\epsilon_{11} + \lambda(\epsilon_{11}+\epsilon_{22}+\epsilon_{33}) & 2\mu\epsilon_{12} & 2\mu\epsilon_{13} \\ 2\mu\epsilon_{21} & 2\mu\epsilon_{22} + \lambda(\epsilon_{11}+\epsilon_{22}+\epsilon_{33}) & 2\mu\epsilon_{23} \\ 2\mu\epsilon_{31} & 2\mu\epsilon_{32} & 2\mu\epsilon_{33} + \lambda(\epsilon_{11}+\epsilon_{22}+\epsilon_{33}) \end{bmatrix} \tag{3.28}$$

Thus, the determination of each element of tensor $\boldsymbol{C}$ can be concluded as

$$\begin{aligned} C_{1111} &= C_{2222} = C_{3333} = 2\mu + \lambda \\ C_{1122} &= C_{1133} = C_{2211} = C_{2233} = C_{3311} = C_{3322} = \lambda \\ C_{1212} &= C_{1313} = C_{2121} = C_{2323} = C_{3131} = C_{3232} = 2\mu \end{aligned} \tag{3.29}$$

The other elements of the tensor $\boldsymbol{C}$ are known to be zero. To simplify and also reduce unnecessary elements, the stress and strain tensor are often sorted in a vector-form instead as

$$\hat{\boldsymbol{\sigma}} = [\sigma_{11}\ \sigma_{22}\ \sigma_{33}\ \sigma_{12}\ \sigma_{23}\ \sigma_{13}] \quad \text{and} \quad \hat{\boldsymbol{\epsilon}} = [\epsilon_{11}\ \epsilon_{22}\ \epsilon_{33}\ 2\epsilon_{12}\ 2\epsilon_{23}\ 2\epsilon_{13}] \tag{3.30}$$

Using the tensors in vector-form, an extra caution must be taken since the conventional coordinate transformation may not valid. Take this into account, the stress tensor is known as,

$$\hat{\boldsymbol{\sigma}} = \hat{\boldsymbol{C}} \cdot \hat{\boldsymbol{\epsilon}} \tag{3.31}$$

For isotropic material, the strain tensor $\boldsymbol{C} := \hat{\boldsymbol{C}} \in \mathbb{R}^{6 \times 6}$ can be simplified as

$$\hat{\boldsymbol{C}} = \frac{E}{(1+\nu)(1-2\nu)} \begin{bmatrix} 1-\nu & \nu & \nu & & & \\ \nu & 1-\nu & \nu & & \boldsymbol{0} & \\ \nu & \nu & 1-\nu & & & \\ & & & \frac{1}{2}-\nu & 0 & 0 \\ & s & \boldsymbol{0} & 0 & \frac{1}{2}-\nu & 0 \\ & & & 0 & 0 & \frac{1}{2}-\nu \end{bmatrix} \tag{3.32}$$

For a further extensive discussion on Cauchy stress tensor, the interested reader is encouraged to consult [15, 26, 101, 102] for the relation between material tensor forms and their matrix quantities.

3.3 Conservation of momentum

The law of conservation of mass states that mass of a closed system will remain constant over time, when the mass to energy conversion is not considered. Therefore, the mass is conserved in reference and current configurations. ρ are the constant material density of a solid body at both reference and current configurations respectively. Thus,

$$m(t) = \int_{\mathcal{S}_\Gamma} \rho(\boldsymbol{x},t)\mathrm{d}v = \int_{\mathcal{B}_\Gamma} \rho(\boldsymbol{X}) J \mathrm{d}V \tag{3.33}$$

The *Newton*'s second law of motion for a continuum or the momentum conservation principle in Equation 3.34 is collectively recognized as the *strong form*. It states that the material time derivative of the linear momentum equals the net force.

$$\dot{\boldsymbol{L}}(t) = \int_{\mathcal{S}_\Gamma} \rho \, \ddot{\boldsymbol{u}} \mathrm{d}v = \int_{\mathcal{S}_\Gamma} \rho \, \boldsymbol{b} \mathrm{d}v + \int_{\partial \mathcal{S}_\Gamma} \boldsymbol{T} \mathrm{d}a \tag{3.34}$$

Where $\rho \boldsymbol{b}$ is body force per unit volume and $\boldsymbol{T}$ is known from Cauchy boundary stress vector acting on the surface $\partial \mathcal{S}_\Gamma$ of a volume. Apply the Cauchy stress vector known from equation 3.20 to equation 3.34 provides

$$\int_{\mathcal{S}_\Gamma} \rho \, \ddot{\boldsymbol{u}} \mathrm{d}v = \int_{\mathcal{S}_\Gamma} \rho \, \boldsymbol{b} \mathrm{d}v + \int_{\partial \mathcal{S}_\Gamma} \boldsymbol{\sigma} \cdot \boldsymbol{n} \mathrm{d}a \tag{3.35}$$

Employing the *Grauß*'s integral theorem, the equation 3.35 can be written as

$$\int_{\mathcal{S}_\Gamma} \rho \, \ddot{\boldsymbol{u}} \mathrm{d}v = \int_{\mathcal{S}_\Gamma} \rho \, \boldsymbol{b} \mathrm{d}v + \int_{\mathcal{S}_\Gamma} \mathrm{div} \boldsymbol{\sigma} \mathrm{d}v \tag{3.36}$$

or

$$\int_{\mathcal{S}_\Gamma} \left(\rho \, \boldsymbol{b} - \rho \, \ddot{\boldsymbol{u}} + \mathrm{div} \boldsymbol{\sigma} \right) \mathrm{d}v = 0 \tag{3.37}$$

Equation 3.37 must be valid for a whole body $\mathcal{S}_\Gamma$, therefore the integrand must equal zero. This assumption points to the local balance of momentum for any particle on the body. Therefore, Equation 3.38 which is known as the governing system of equations is obtained.

$$\rho \left(\boldsymbol{b} - \ddot{\boldsymbol{u}} \right) + \mathrm{div} \boldsymbol{\sigma} = 0 \tag{3.38}$$

or index-wise,

$$\rho \left(b_i - \ddot{u}_i \right) + \frac{\partial \sigma_i}{\partial x_j} = 0 \tag{3.39}$$

3.4 Boundary value problems

The purpose of applying boundary values to the governing system of equations is to solve the equations with prescribed support displacements, tension forces and the

3.4 Boundary value problems

geometry alternation due to cut.

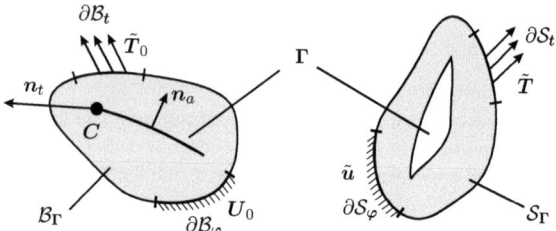

Figure 3.3: Boundary value problem of a solid body in reference $\mathcal{B}_\Gamma$ and current $\mathcal{S}_\Gamma$ configurations

The main challenge of modeling of an incision process lies by how the material structural changes are properly treated as the boundary conditions. To begin with, Figure 3.3 is referred. The same solid body $\mathcal{B}$ has the outer boundary values $\partial \mathcal{S}$ in its current configuration of domain $\mathcal{S}_\Gamma$. $\partial \mathcal{S}$ are given by the all known boundary conditions including where the cut area exists.

$$\partial \mathcal{S} = \partial \mathcal{S}_\varphi \cup \partial \mathcal{S}_t \qquad (3.40)$$

Analogously, both boundaries exhibits no intersection. Thus,

$$\partial \mathcal{S}_\varphi \cap \partial \mathcal{S}_t = \emptyset \qquad (3.41)$$

Prescribed displacements $\partial \mathcal{S}_\varphi$ are known as *Dirichtlet* boundary conditions or essential boundary conditions. The prescribed displacements in current configuration are denoted by $\tilde{\boldsymbol{u}}$

In contrast, prescribed loads $\partial \mathcal{S}_t$ on the surface of the solid body and the cut are referred as *Neumann* boundary conditions or natural boundary conditions. They are assumed automatically on all remaining boundaries. The prescribed loads in current configuration are given by the traction force vector $\tilde{\boldsymbol{T}}$. In addition, the traction caused by the cutting tool to the solid body is described as part of prescribed loads. However, the inner boundary condition of both faces surrounding the cut Γ recalls no surface traction. Thus,

$$\boldsymbol{\sigma} \cdot \boldsymbol{n}_a = \tilde{\boldsymbol{T}}(\boldsymbol{x}) = 0 \quad \text{where} \quad \boldsymbol{x} \in \partial \Gamma \qquad (3.42)$$

3.4.1 Initial value problems

The initial value problems considers all boundary conditions $\partial \mathcal{B}$ correspond to the state of the solid body in its reference configuration $\mathcal{B}_-$. In technical mechanics, the initial state often correspond to the reference time $t = 0$. Since the incision employs the material structural changes Ξ_t and causes the changes in the reference configuration, the context of reference is therefore confusing due to the reference configuration is time-dependent according to the material structural changes [36].

Despite the extensive use of the classical definition of the initial boundary values in this work, they are defined in a different manner. The current configuration in this work is clearly denoted by the configuration which the solid body occupies after the material structural changes. Their boundary values are given similarly to the ones of the current configuration as $\partial \mathcal{B} = \partial \mathcal{B}_\varphi \cup \partial \mathcal{B}_t$. The time-dependency of the reference configuration forces the boundary values to be remapped as discussed in section 3.5

The actual prescribed displacements and loads in reference configuration are required for the governing system of equations to be solvable. Additionally, if the dynamic behavior is as of interest, the displacement velocity must be given as well. The initial boundary value problems can be summarized as

$$U(X, t=0) = U_0 \tag{3.43}$$
$$\dot{U}(X, t=0) = \dot{U}_0 \tag{3.44}$$
$$\tilde{T}(X, t=0) = \tilde{T}_0 \quad \text{and} \tag{3.45}$$
$$\tilde{T}(X, t=0) = 0 \quad \text{where} \quad X \in \Gamma \tag{3.46}$$

3.5 Remapping of the governing system of equations

The governing system of equations 3.38 illustrates the dynamic deformation of a solid body with respect to its current geometry, body force and the surface traction applied in its current configuration. On the other hand, domain $\mathcal{B}_\Gamma :=$ $\mathcal{B} \setminus \{\Gamma(t) \cup \partial \Gamma(t)\}$ changes over time as the cut $\Gamma(t)$ evolves. Depending on how the material structural changes Ξ_t occur – or in context of this work, the type of cut – the governing system of equations must be reconsidered as follow.

- Figure 3.4 refers to an incision process which employs a sharp cutting instrument such as a medical scalpel or a crack demonstrating only a split in structure. Ξ_t maps every material P on the cut in Ω_Γ and defy the geometry of the solid body in reference configuration with X. This causes materials originate on domain $\mathcal{B}_\Gamma$ to be separated on the cut surface $\partial \Gamma$. The material mass distribution is required to be altered according to Ξ_t. The conservation of mass from equation 3.33 exhibits no change to the total mass in the solid body and remain constant ($\bm{m}(t) = \bm{m}_0$). The displacement $\bm{u}$ of current configuration $\bm{x}$ can be obtained by solving the governing system of equations.

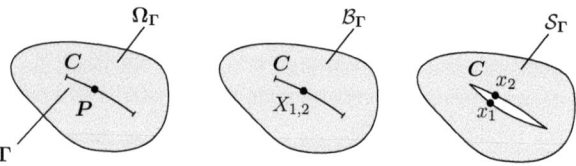

Figure 3.4: Mapping change due to incision or split in structure

- In case of Figure 3.5, a series of incision is done to enclose part of the solid body. It results in a material removal. The total amount of mass is therefore time-dependent and preserved only temporarily during mapping between reference and current configuration. $\mathcal{B}_\Gamma$ remains a closed system for that particular time interval t. Hence, in case of material removal, the governing system of equations is required to be reconsidered not only the mass distribution of the materials on the cut surface but also the total mass in the solid body.

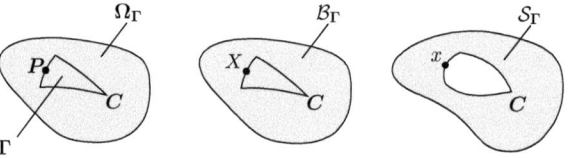

Figure 3.5: Mapping change and material removal

3.6 Weak form of conservation of momentum

The conservation of momentum cannot be discretized directly by the finite element method. To solve this problem, the *weak form* of the conservation of momentum and the traction boundary conditions which is known as the *strong form* must be derived using the principle of virtual work.

3.6 Weak form of conservation of momentum

3.6.1 Strong to weak form

A weak form will now be developed for the conservation of momentum introduced in Equation 3.34 and the traction boundary conditions mentioned in the section 3.4. For this purpose, the trial function $u(X,t)$ is employed to satisfy all displacement boundary conditions and to be smooth enough so that all derivatives in the conservation of momentum equation are well defined. The test function $\delta u(X)$ are assumed to be smooth enough so that all of the following steps are well defined and to vanish on the prescribed displacement boundary.

The weak form is obtained by taking the product of the conservation of momentum equation with the test function and integrating over the domain [15]. The integral of this product must be zero. For that in the reference configuration,

$$\int_{\mathcal{B}_r} [\rho(\boldsymbol{b} - \ddot{\boldsymbol{u}}) + \nabla_x \boldsymbol{\sigma}] \delta \boldsymbol{u} \, dV = 0 \tag{3.47}$$

or with equation 3.39, the equation above can be written index-wise as

$$\int_{\mathcal{B}_r} \left[\rho(b_i - \ddot{u}_i) + \frac{\partial \sigma_{ik}}{\partial x_k} \right] \delta u_j dV = 0 \tag{3.48}$$

with the product rule known as

$$\frac{\partial a(x)}{\partial x} b(x) = \frac{\partial (a(x)b(x))}{\partial x} - a(x)\frac{\partial b(x)}{\partial x} \tag{3.49}$$

By-part integration of the stress term by employing the Grauß's integral theorem yields

$$-\int_{\mathcal{B}_r} \frac{\partial \delta u_j}{\partial x_k} \sigma_{ij} dV + \int_{\mathcal{B}_r} \frac{\partial (\delta u_j \sigma_{ik})}{\partial x_k} dV + \int_{\mathcal{B}_r} \rho(b_i - \ddot{u}_i) \delta u_j dV = 0 \tag{3.50}$$

Using the divergence theorem, the surface traction is derived as

$$\int_{\mathcal{B}_r} \frac{\partial (\delta u_j \sigma_{ik})}{\partial x_k} dV = \int_{\partial \mathcal{B}_r} \delta u_j \sigma_{ik} n_k dA = \int_{\partial \mathcal{B}_r} \delta u_j T_i dA \tag{3.51}$$

Substitution of this surface traction in equation 3.50 provides

$$-\int_{\mathcal{B}_r} \frac{\partial \delta u_j}{\partial x_k} \sigma_{ik} dV + \int_{\partial \mathcal{B}_r} \delta u_j T_i dA + \int_{\mathcal{B}_r} (\delta u_j b_i - \delta u_j \rho \ddot{u}_i) dV = 0 \tag{3.52}$$

The term $\int_{\partial \mathcal{B}_r} \delta u_j T_i dA$ are applied to all initial Dirichlet boundary conditions or the essential boundary conditions known from section 3.4 as $\tilde{T}_0(X)$. Thus the weak-

3.6 Weak form of conservation of momentum

form of the governing system of equation is

$$\int_{\mathcal{B}_\Gamma} \delta u_j \rho b_i \mathrm{d}V - \int_{\partial\mathcal{B}_\Gamma} \delta u_j \tilde{T}_0 \mathrm{d}A + \int_{\mathcal{B}_\Gamma} \frac{\partial \delta u_j}{\partial x_k} \sigma_{ik} \mathrm{d}V + \int_{\mathcal{B}_\Gamma} \delta u_j \rho \ddot{u}_i \mathrm{d}V = 0 \qquad (3.53)$$

or

$$\int_{\mathcal{B}_\Gamma} \delta \boldsymbol{u} \rho \boldsymbol{b} \mathrm{d}V - \int_{\partial\mathcal{B}_\Gamma} \delta \boldsymbol{u} \tilde{\boldsymbol{T}}_0 \mathrm{d}A + \int_{\mathcal{B}_\Gamma} \frac{\partial \delta \boldsymbol{u}}{\partial \boldsymbol{x}} \boldsymbol{\sigma} \mathrm{d}V + \int_{\mathcal{B}_\Gamma} \delta \boldsymbol{u} \rho \ddot{\boldsymbol{u}} \mathrm{d}V = 0 \qquad (3.54)$$

3.6.2 Smoothness of the test function and the trial function

In classical derivations of the weak form, all functions appeared in the strong form are assumed to be continuous. For the conservation of momentum from equation 3.38 to be well defined in a classical sense, the product of the nominal stress and the initial area must be continuously differentiable, i.e. C^1. Otherwise the first derivative would exhibits discontinuities [15]. The trial function $\delta\boldsymbol{u}$ needs to satisfy all displacement boundary conditions. These conditions on the trial displacement are indicated symbolically by

$$\boldsymbol{u}(\boldsymbol{X},t) \in \mathcal{U} \qquad (3.55)$$

where $\mathcal{U} = \left\{ \boldsymbol{u}(\boldsymbol{X},t) \mid \boldsymbol{u}(\boldsymbol{X},t) \in {}^0\mathcal{C}(\boldsymbol{X}), \boldsymbol{u} = \boldsymbol{U}_0 \text{ on } \partial\mathcal{B}_\varphi \right\}$

Displacement fields which satisfy the condition pointed by equation 3.55, such as the displacement fields which are in $\mathcal{U}$, are called *kinematically admissible*. The test function $\delta\boldsymbol{u}$ is not time-dependent. Moreover, the test function is required to be C^0 and to vanish on displacement boundaries, such as

$$\delta\boldsymbol{u}(\boldsymbol{X},t) \in \mathcal{U}_0 \qquad (3.56)$$

where $\mathcal{U}_0 = \left\{ \delta\boldsymbol{u}(\boldsymbol{X},t) \mid \boldsymbol{u}(\boldsymbol{X},t) \in \mathcal{C}^0(\boldsymbol{X}), \delta\boldsymbol{u} = 0 \text{ on } \partial\mathcal{B}_\varphi \right\}$

The prefix δ is denoted for all variables which are test functions and for variables which are functions of the test functions.

3.6.3 Principle of virtual work

For the purpose of obtaining a methodical procedure for developing the finite element equations, the virtual works will be defined according to the type of work they represent. Each of the terms in the weak form represent the work occurred due to the test function $\delta\boldsymbol{u}$, which is often called the virtual displacement to indicate that it is actually the virtual and not an actual displacement. The principle of virtual work says

$$\delta W(\delta\boldsymbol{u},\boldsymbol{u}) \equiv \delta W^{int} - \delta W^{ext} + \delta W^{kin} = 0 \qquad (3.57)$$

3.6 Weak form of conservation of momentum

The virtual work of the body force and the traction corresponds to the first and second terms in Equation 3.54 is called the virtual external work since it is the result of the external loads and given by

$$\delta W^{ext} = \int_{\mathcal{B}_r} \delta \boldsymbol{u} \rho \mathbf{b} \, dV - \int_{\partial \mathcal{B}_r} \delta \boldsymbol{u} \tilde{\boldsymbol{T}}_0 \, dA \qquad (3.58)$$

The third term of Equation 3.54 reflects the virtual internal work as a consequence of the stresses in the material

$$\delta W^{int} = \int_{\mathcal{B}_r} \frac{\partial \delta \boldsymbol{u}}{\partial \boldsymbol{x}} \boldsymbol{\sigma} \, dV \qquad (3.59)$$

while the last term can be considered as a body force which acts in their direction opposite to the acceleration, i.e. a *d'Alembert* force. In context of this work, the term is recognized as the virtual kinetic works, thus

$$\delta W^{kin} = \int_{\mathcal{B}_r} \delta \boldsymbol{u} \rho \ddot{\boldsymbol{u}} \, dV \qquad (3.60)$$

The benefit of the virtual work in this viewpoint is to obtain the weak form for further discretization in the finite element approximation. Thus, the steps from multiplying the equation with the test function and performing various manipulations can actually be avoided. The virtual work scheme is useful for memorizing the weak form due to its representativeness to the actual physical property of a closed system [15]. From mathematical perspective, the test function $\delta \boldsymbol{u}$ is not required to be a virtual displacements. It is allowed to be any test function which satisfies continuity conditions and vanish on the displacement boundaries as described by equation 3.56.

4 Finite element approach

The Finite Element Method or FEM is well known for its suitability and scalability of modeling a deformable body. FEM gains a major popularity in the real-time haptic and surgical simulation. It is also a suitable tool to discrete the governing system of equations obtained from chapter 3 for contact model implementation. Using FEM A similar approaches can be seen in the works of demonstrating different types of contact mechanics for a telepresence application [71, 100].

Both analysts and developers of FEM-based simulation should understand the fundamental concepts of finite element analysis thoroughly. Without an understanding of the implication and meaning of the physical interpretation, FEM approximation may be not be usable. Therefore, this chapter is devoted to an extensive discussion on the discretization of the weak form of the governing system of equations obtained in section 3 and how the spatial discretization of FEM principle can be applied to an individual material volume or element. This section will also addresses the remapping problem between configurations which occurs due to a rapid change in the geometry of a soft body when an incision process is carried out.

4.1 Spatial discretization

The approximation using finite elements begins with the spatial discretization of a problem domain into finite numbers of units or elements. For a Lagrangian mesh, each element processes nodes mapped to fixed material points or particles. The arrangement of the elements is called the topology indexing of nodal degree-of-freedoms (DoFs) in the governing system of equations. Thus, an element can be identified by the index of the DoFs it uniquely occupies. In Figure 4.1(a) the solid body $\mathcal{B}$ is discretized without exhibiting of a cut. The discrete body is a representative of the considered deformation problem governing by the system of equations 3.38 and 3.54. Figure 4.1(b) depicts the same solid body $\mathcal{B}$ with a cut, the elements in the discrete representative body are arranged so that the elements match the geometry of the cut. Both Figures demonstrate that the representative bodies do not match the actual domains neither with or without a cut exactly. Thus, the accuracy of the approximation obtained from the model depends strongly on how a solid body is discretized since the beginning.

Also, if the solid body initially processed no cut in its body, a re-discretization is necessary so that the geometry of the cut is presented properly. The topology is updated so that the discrete body still represents the actual body geometry. Depend on the cut, the topology change may increase the number of elements required in

4.1 Spatial discretization

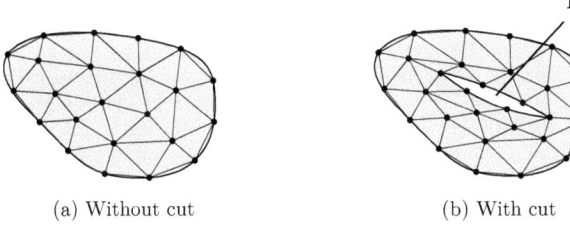

(a) Without cut (b) With cut

Figure 4.1: Spatial discrete solid body

the re-discretized body, hence the number of DoF increases. The discussion on the spatial re-discretization for a cut will be carried on in section 4.4.1.

Figure 4.2 depicts the same solid body $\mathcal{B}$ occupying at a specific time t the domain of current configuration $\mathcal{S}_\Gamma$. The solid body is now represented by set of subdomains Ω_e. The amount of subdomains is known as the number of elements n_{ele}. Since the solid body occupies a domain $\mathcal{B}_\Gamma := \mathcal{B} \setminus \{\Gamma(t) \cup \partial\Gamma(t)\}$ which excludes the cut $\Gamma(t)$, the domain of current configuration $\mathcal{S}_\Gamma$ inherits piece-wise mapping and represents the similar domain in a different configuration.

$$\mathcal{S}_\Gamma \approx \sum_{e=1}^{n_{ele}} \Omega_e \tag{4.1}$$

Similarly, the boundary of the solid body at a specific current time is divided into

$$\partial\mathcal{S}_\Gamma \approx \sum_{e=1}^{n_{ele}} \partial\Omega_e \tag{4.2}$$

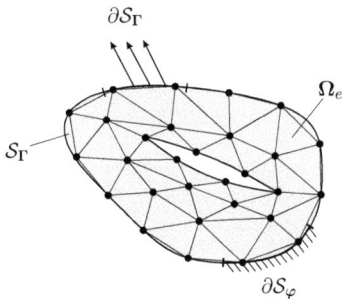

Figure 4.2: Discretization to the solid body into number of subdomains Ω_e

4.2 Discrete governing system of equations

The discrete equations for a finite element model are obtained from the principle of virtual work by using finite element interpolation for the test and trial functions. It employs the approximated solution on the nodes of each element to describe the motion of a solid body. The discretization is established in the initial configuration $\mathcal{B}_\Gamma$ using isoparametric elements to interpolate the initial geometry in terms of the particle denoted by $\boldsymbol{X}_j$ which defining the initial position of nodes in an element as

$$\boldsymbol{X} \approx \sum_{j=1}^{n} N_j(\boldsymbol{X})\boldsymbol{X}_j \tag{4.3}$$

where $N_j(\boldsymbol{X})$ are a shape function and n denotes the number of nodes per element. The shape function is be chosen to comply with the Kronecker Delta property

$$N_j(X_i) := \delta_{ij} = \begin{cases} 0; & i \neq j \\ 1; & i = j \end{cases} \tag{4.4}$$

and the Partition of Unity

$$\sum_{j=1}^{n} N_j(\boldsymbol{X}) = 1 \tag{4.5}$$

As for a Lagrangian mesh, during the motion, nodes and elements are permanently attached to the particles with which they were initially associated. Accordingly, the subsequent motion can be described in terms of the current configuration $\boldsymbol{x}_j(\boldsymbol{X},t)$ of a nodal particle.

$$\boldsymbol{x} = \sum_{j=1}^{n} N_j(\boldsymbol{X})\boldsymbol{x}_j \tag{4.6}$$

In consistent with their kinematic derived in equation 3.7, the displacement $\boldsymbol{u}$ is interpolated as

$$\boldsymbol{u} = \sum_{j=1}^{n} N_j(\boldsymbol{X})\boldsymbol{u}_j \tag{4.7}$$

The test function known as virtual displacement $\delta\boldsymbol{u}$ is required to satisfy the linear independence which is necessary to solve the discrete governing system of equations. The shape functions are also used for the test function $\delta\boldsymbol{u}$ in the weak form

$$\delta\boldsymbol{u} = \sum_{j=1}^{n} N_j(\boldsymbol{X})\delta\boldsymbol{u}_j \tag{4.8}$$

4.2 Discrete governing system of equations

Analogously, the nodal displacement velocity and acceleration fields are obtained as

$$\dot{u} = \sum_{j=1}^{n} N_j(X)\dot{u}_j \quad \text{and} \quad \ddot{u} = \sum_{j=1}^{n} N_j(X)\ddot{u}_j \qquad (4.9)$$

For a three-dimensional configuration, the shape function can be arranged in a matrix as

$$N = \begin{bmatrix} N_1 & 0 & 0 & \cdots & N_n & 0 & 0 \\ 0 & N_1 & 0 & \cdots & 0 & N_n & 0 \\ 0 & 0 & N_1 & \cdots & 0 & 0 & N_n \end{bmatrix} \qquad (4.10)$$

The weak form of conservation of momentum derived in equation 3.54 was obtained to associate with the discretization of the FEM. From Equation 3.24 and 4.7, the infinitesimal strain tensor is approximated as

$$\epsilon_{ij} = \frac{1}{2}\left(\frac{\partial u_i}{\partial x_j} + \frac{\partial u_j}{\partial x_i}\right) = \frac{1}{2}\left(\frac{\partial N_{ik}u_k}{\partial x_j} + \frac{\partial N_{jk}u_k}{\partial x_i}\right) \qquad (4.11)$$
$$= \frac{1}{2}\left(\frac{\partial N_{ik}}{\partial x_j} + \frac{\partial N_{jk}}{\partial x_i}\right)u_k$$

The strain-displacement matrix $B \in \mathbb{R}^{6 \times 3n}$

$$\hat{\epsilon} = B \cdot u \qquad (4.12)$$

with

$$B = \begin{bmatrix} \frac{\partial N_1}{\partial x_1} & 0 & 0 & \cdots & \frac{\partial N_n}{\partial x_1} & 0 & 0 \\ 0 & \frac{\partial N_1}{\partial x_2} & 0 & \cdots & 0 & \frac{\partial N_n}{\partial x_2} & 0 \\ 0 & 0 & \frac{\partial N_1}{\partial x_3} & \cdots & 0 & 0 & \frac{\partial N_n}{\partial x_3} \\ \frac{\partial N_1}{\partial x_2} & \frac{\partial N_1}{\partial x_1} & 0 & \cdots & \frac{\partial N_n}{\partial x_2} & \frac{\partial N_n}{\partial x_1} & 0 \\ 0 & \frac{\partial N_1}{\partial x_3} & \frac{\partial N_1}{\partial x_2} & \cdots & 0 & \frac{\partial N_n}{\partial x_3} & \frac{\partial N_n}{\partial x_2} \\ \frac{\partial N_1}{\partial x_3} & 0 & \frac{\partial N_1}{\partial x_1} & \cdots & \frac{\partial N_n}{\partial x_3} & 0 & \frac{\partial N_n}{\partial x_1} \end{bmatrix} \qquad (4.13)$$

is introduced. From Equation 3.32, substituting the strain tensor $\hat{\epsilon}$ in the equation 4.12 yields,

$$\hat{\sigma} = \hat{C} \cdot \hat{\epsilon} = \hat{C} \cdot B \cdot u \qquad (4.14)$$

The weak form from Equation 3.54 is applied to an element Ω_e with the discretized u and the shape function to obtain

$$\left[\int_{\Omega_e} N^T \rho b \mathrm{d}V - \int_{\partial \Omega_e} N^T \tilde{T}_0 \mathrm{d}A + \int_{\Omega_e} B^T \hat{C} B \mathrm{d}V u + \int_{\Omega_e} N^T \rho N \mathrm{d}V \ddot{u}\right] \delta u = 0 \qquad (4.15)$$

which introduces the following physical terms of an element

$$k^{(e)} = \int_{\Omega_e} B^T \hat{C} B \, dV \quad \in \mathbb{R}^{3n \times 3n} \quad \text{Stiffness matrix} \quad (4.16)$$

$$m^{(e)} = \int_{\Omega_e} N^T \rho N \, dV \quad \in \mathbb{R}^{3n \times 3n} \quad \text{Mass matrix} \quad (4.17)$$

$$f_{\Omega}^{(e)} = \int_{\Omega_e} N^T \rho b \, dV \quad \in \mathbb{R}^{3n} \quad \text{Body volume force vector} \quad (4.18)$$

$$f_{\partial\Omega}^{(e)} = \int_{\partial\Omega_e} N^T \tilde{T}_0 \, dA \quad \in \mathbb{R}^{3n} \quad \text{Surface traction vector} \quad (4.19)$$

The element damping matrix is a material property required for the dynamic simulation of a soft body in this work. It may be possible for the global damping matrix to be found by considering the total energy lost of the whole solid body [13] but not for an element. Nevertheless, the calculated global damping matrix may be dependent to eigenfrequency of the solid body. Therefore, a decoupling element damping similar to element mass and stiffness matrices can not be obtained. Because of this complexity of damping modeling, a common type of damping called Rayleigh damping is often employed in the linear and nonlinear incremental analysis of structures [12, 92]. Rayleigh damping assumes that the damping matrix is decoupled and proportional to the mass and stiffness matrices in the form of

$$d^{(e)} = a\, m^{(e)} + b\, k^{(e)} \quad \in \mathbb{R}^{3n \times 3n} \quad \text{Damping matrix} \quad (4.20)$$

The coefficients a and b are selectable to obtain a desire damping value. Using Rayleigh damping yields another benefit necessary for vibration analysis. This implementation also allows the damping to be increased as the eigenfrequency increases when a stiffer material is employed [23]. In this work, this approach is also adopted. The discrete dynamic governing equation for an element Ω_e is therefore obtained in the following form.

$$k^{(e)} \cdot u + d^{(e)} \cdot \dot{u} = f_{\Omega}^{(e)} + f_{\partial\Omega}^{(e)} - m^{(e)} \cdot \ddot{u} \quad (4.21)$$

4.3 Element coordinates

FEM is usually developed with shape functions expressed in terms of parent element coordinates, often called element coordinates for brevity. The element coordinates can be considered an alternative set of material coordinates in a well-adopted Lagrangian mesh. Therefore, the shape functions in terms of element coordinates is intrinsically equivalent to expressing a term in material coordinates. The shape of the parent domain depends on the shape and the dimension of the element used to discrete a solid body. The element coordinate is denoted by ζ in tensor notation. The map $X = x(\zeta, 0)$ corresponds to the reference or initial configuration, while

4.3 Element coordinates

$\boldsymbol{x} = \boldsymbol{x}(\zeta,t)$ refers to the current configuration which is time-dependent. Thus, there is valid a chain rule describing the shape functions in element coordinate as

$$\frac{\partial N_i}{\partial \boldsymbol{x}} = \frac{\partial N_i}{\partial \boldsymbol{\zeta}} \cdot \frac{\partial \boldsymbol{\zeta}}{\partial \boldsymbol{x}} = \frac{\partial N_i}{\partial \boldsymbol{\zeta}} \cdot \boldsymbol{J}^{-1} \quad \text{or} \quad \frac{\partial N_i}{\partial x_j} = \frac{\partial N_i}{\partial \zeta_k} \frac{\partial \zeta_k}{\partial x_j} = \frac{\partial N_i}{\partial \zeta_k} \left[\boldsymbol{J}^{-1}\right]_{kj} \quad (4.22)$$

Because the determinant of the Jacobian refers to the volume change of an element with respect to the element coordinate, $\det(\boldsymbol{J}) > 0$. If $\det(\boldsymbol{J}) \leq 0$ at any time, it means that the material density $\rho \leq 0$, which is physically impossible [15]. Therefore, the Jacobian must be invertible.

In this work, solid bodies are all discretized using tetrahedral elements. As depicted in Figure 4.3, a tetrahedral element occupies a three-dimensional space with 4 nodes and has 12 degrees of freedom in total. It has 4 shape functions, which are $N_1 = \zeta_1$, $N_2 = \zeta_2$, $N_3 = \zeta_3$ and $N_4 = 1 - \zeta_1 - \zeta_2 - \zeta_3$.

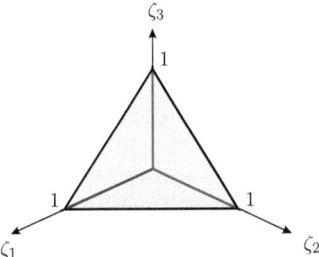

Figure 4.3: A tetrahedral element

For a linear Finite Element approximation, the derivative of the shape function with respect to the element coordinate is constant as follow

$$\frac{\partial \boldsymbol{N}}{\partial \boldsymbol{\zeta}} = \begin{bmatrix} 1 & 0 & 0 \\ 0 & 1 & 0 \\ 0 & 0 & 1 \\ -1 & -1 & -1 \end{bmatrix} \quad (4.23)$$

With these properties, the strain-displacement matrix $\boldsymbol{B}$ from equation 4.12 can be written in element coordinate as well as the integrand $\boldsymbol{B}\hat{\boldsymbol{C}}\boldsymbol{B}$. Hence, the element stiffness $\boldsymbol{k}^{(e)}$ can be written in element coordinate specific for tetrahedral element as

$$\boldsymbol{k}^{(e)} = \int_{\Omega_e} \boldsymbol{B}^T \hat{\boldsymbol{C}} \boldsymbol{B} \mathrm{d}V = \int_0^1 \int_0^{1-\zeta_3} \int_0^{1-\zeta_2-\zeta_3} \boldsymbol{B}^T \hat{\boldsymbol{C}} \boldsymbol{B} \det \boldsymbol{J} \ \mathrm{d}\zeta_1 \mathrm{d}\zeta_2 \mathrm{d}\zeta_3 \quad (4.24)$$

In a similar manner, the element mass matrix $m^{(e)}$ in equation 4.17 is found as

$$m^{(e)} = \int_{\Omega_e} N^T \rho N \mathrm{d}V = \int_0^1 \int_0^{1-\zeta_3} \int_0^{1-\zeta_2-\zeta_3} N^T \rho N \det J \, \mathrm{d}\zeta_1 \mathrm{d}\zeta_2 \mathrm{d}\zeta_3 \quad (4.25)$$

4.4 Assembly and remeshing

4.4.1 Assembly process

In Lagrangian mesh, the nodes and elements move with the material. Boundaries and interfaces coincide with element edges, so that their treatment is simplified. Quadrature points also move with the material, thus constitutive equations are always evaluated at the same material points, which is advantageous for history-dependent materials and boundary conditions [15].

Each discrete element represents a certain amount of material in the solid body which is decided during the discretization. As discussed in section 4.1, an element occupies a set of individual nodes called topology. The elements whose at least one node share the same index of DoF are neighbor in Lagrangian mesh. These element are coupled with each other. The assembly process is referred here by the unification operator $\bigcup$. It sorts a variable of neighbor elements together using their unique indexes. Hence, nodal mass of each element can be assembled to together to provide a global mass matrix

$$M = \bigcup_{e=1}^{n_{ele}} m^{(e)} \quad (4.26)$$

the global stiffness matrix

$$K = \bigcup_{e=1}^{n_{ele}} k^{(e)} \quad (4.27)$$

and the global damping matrix

$$D = \bigcup_{e=1}^{n_{ele}} d^{(e)} \quad (4.28)$$

The global element body force and the global surface traction force can also assembled together as

$$F^{ext} = \bigcup_{e=1}^{n_{ele}} f_{\partial\Omega}^{(e)} + \bigcup_{e=1}^{n_{ele}} f_{\Omega}^{(e)} \quad (4.29)$$

As result, a discrete global dynamic system of equations is obtained in Equation 4.30

4.4 Assembly and remeshing

with the global nodal displacement denoted by the vector U.

$$F^{ext} = M\ddot{U} + D\dot{U} + KU \quad (4.30)$$

4.4.2 Remeshing

As mentioned earlier, Lagrangian meshes let the nodes and elements attached permanently to the material points. The discussion in section 3.5 is the necessity of remapping $\Xi_t : \Omega_\Gamma \to \mathcal{B}_\Gamma$ to preserve the validity of the governing system of equations. For an incision simulation which the process contains a constant evolve of a material discontinuity in a solid body, the remeshing process is compulsory for the mesh to reasonably correspond to the actual geometry. In general, remeshing processes repeats the spatial discretization to a solid body into, so that the present geometry of the solid body is adequately represented.

The remeshing algorithm in this work is done by allowing the simulated object to be cut in a vertical plane. Additionally, the scalpel is appointed to pass through a series of nodes connecting a set of tetrahedral elements, as shown in Figure 4.4. The employed algorithm is required to demonstrate the increasing of the DoFs around the opened surface. This methodology can be realized because the simulated object is discretized so that the tetrahedral elements are uniformly arranged. A node is pointed by a unique index number which is one of local properties of a tetrahedral element. To repeat the use of the same node number with no additional argument can be implied that two or more elements are occupying the same node and the cut does not exist between them. For this reason, the original nodes shall be multiplied by new additional nodes. Hence the total DoFs of the solid body will be generally increased due to the expansion of the surface.

Neumann and Dirichlet boundary conditions from the original mesh can be further adopted without any alternation, if the cut does not occur on the area where the Neumann and Dirichlet boundary conditions are constantly applied during the simulation.

A remeshing process is required to be initiated each time the geometry change is exhibited. The change of DoF can be implied mathematically that the topology of the governing system of equations 3.38 is not preserved and the actual solid body is not identical to the original ones. Therefore, the assembly process must be called right after the remeshing process to reassembly a new governing system of equations. There are remeshing techniques developed and optimized for each simulation problem. Interesting algorithms can be found in FEM-remeshing for crack growth of TRADEGARD [93] and the tissue resection using delayed updates in a tetrahedral mesh shown done by KUNDU in [49].

4.5 Dynamic simulation

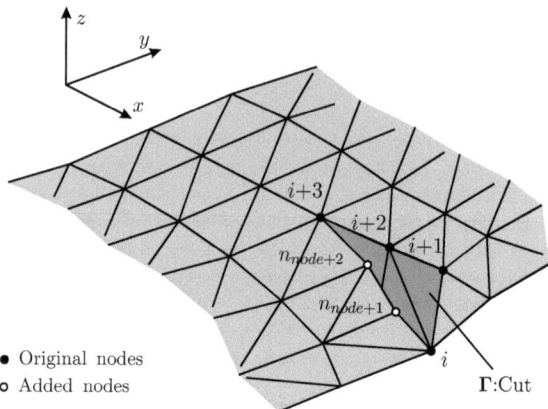

Figure 4.4: Remeshing and updating in an uniform tetrahedral mesh

4.5 Dynamic simulation

In order to calculate the nodal displacement of the solid body, the discrete governing system of equations 4.30 must be integrated or solved. A proper integration method is essential for the accurate results to be obtained, while minimum computing resources are required to stay efficient. This is critical for a real-time simulation application and should be considered thoroughly. This section will discuss on how *Newmark Implicit Integration* is derived as well as the solver of the linear algebraic equations.

4.5.1 Newmark implicit integration

In this work, the well-adopted Newmark Implicit Integration is adopted for its exquisite ability to maintain the stability of the simulation [61]. Unlike the *Central Difference Method* which employs the quadratic polynomial interpolation [68], the Newmark method uses a local different cubic interpolation instead.

$$U(t) = b_0 + b_1 t + b_2 t^2 + b_3 t^3 \tag{4.31}$$

In order to find the coefficient vectors b_i in eqaution 4.31, four vectorial equations are required. The polynomial should exactly interpolate the current displacement U and its velocity $\dot{U}$ at time step t_n as

$$U(t_n) = U_n \quad \text{and} \quad \dot{U}(t_n) = \dot{U}_n \tag{4.32}$$

4.5 Dynamic simulation

Therefore, equation 4.31 is written in explicit form as

$$U(t) = U_n + \dot{U}_n(t - t_n) + b_2(t - t_n)^2 + b_3(t - t_n)^3 \qquad (4.33)$$

To determine the unknown displacement and its velocity at time $t_{n+1} = t_n + \Delta t_n$, the following approaches are adopted from Burnett [20] by interpolate the displacement of time step t_{n+1}

$$U(t_{n+1}) = U_{n+1} = U_n + \dot{U}_n \Delta t_n \frac{1}{2} \left((1 - 2\beta) \ddot{U}_n + 2\beta \ddot{U}_{n+1} \right) \Delta t_n^2 \qquad (4.34)$$

$$\dot{U}(t_{n+1}) = \dot{U}_{n+1} = \dot{U}_n + \left((1 - \gamma) \ddot{U}_n + \gamma \ddot{U}_{n+1} \right) \Delta t_n \qquad (4.35)$$

However, displacements and their velocities in both equations can not be calculated because the acceleration $\ddot{U}_{n+1}$ is still unknown. The following step of formulation employs an *implicit* method which requires the information just only to the time of consideration. Thus, the acceleration from equation 4.34 is found as

$$\ddot{U}_{n+1} = \frac{U_{n+1} - U_n - \dot{U}_n \Delta t_n}{\beta \Delta t_n^2} - \left(\frac{1}{2\beta} - 1 \right) \ddot{U}_n \qquad (4.36)$$

This equation can be put into velocity in equation 4.35 to obtain

$$\dot{U}_{n+1} = \dot{U}_n + (1 - \gamma) \Delta t_n \ddot{U}_n + \gamma \Delta t_n \ddot{U}_{n+1}$$
$$= \dot{U}_n + (1 - \gamma) \Delta t_n \ddot{U}_n +$$
$$\gamma \Delta t_n \left(\frac{U_{n+1} - U_n - \dot{U}_n \Delta t_n}{\beta \Delta t_n^2} - \left(\frac{1}{2\beta} - 1 \right) \ddot{U}_n \right) \qquad (4.37)$$

Substitute the acceleration in 4.36 and velocity in 4.37 in the linear dynamic governing system of equations 4.30 at time t_{n+1} and obtain

$$M\ddot{U}_{n+1} + D\dot{U}_{n+1} + KU_{n+1} = F_{n+1}^{ext} \qquad (4.38)$$

which is known in a recursive form as

$$K_n^* U_{n+1} = F_{n+1}^{ext} + F_n^{int} \qquad (4.39)$$

with

$$K_n^* = \left[\frac{1}{\beta \Delta t_n^2} M + \frac{\gamma}{\beta \Delta t_n} D + K \right] \qquad (4.40)$$

$$F_n^{int} = \left[\frac{1}{\beta \Delta t_n^2} M + \frac{\gamma}{\beta \Delta t_n} D \right] U_n + \left[\frac{1}{\beta \Delta t_n} M + \left(\frac{\gamma}{\beta} - 1 \right) D \right] \dot{U}_n +$$
$$\left[\left(\frac{1}{2\beta} - 1 \right) M + \left(\frac{\gamma}{2\beta} - 1 \right) \Delta t_n D \right] \ddot{U}_n \qquad (4.41)$$

Using Newmark method exhibits the use of parameters β and γ, which stay in the range of $0 < \beta \leq 0.5$ and $0 \leq \gamma \leq 1$. To choose proper values of these parameters,

4.5 Dynamic simulation

the following consideration must be made

- In general the Newmark integration by a linear system remains stable unconditionally for $\gamma \geq 0.5$ and $\beta \geq (\gamma + 0.5)^2/4$.

- For $\gamma = 2\beta$, the cubic interpolation will be comparable to the quadratic interpolation and results in a constant acceleration.

- For $\beta = 0.25$ and $\gamma = 0.5$, the integration maintains unconditionally stable. This choice satisfies the trapezoidal rule which is also known as average acceleration method. It was primary values proposed by Newmark. These values are majority in most of applications.

- With $\beta = 1/6$ and $\gamma = 0.5$, the method is comparable to linear acceleration methods and is only conditionally stable depending on the system.

- If $\beta = 0$ and $\gamma = 0.5$, Newmark integration is identical to the Central Difference method. This choice of parameters leads to a problem that the simulation time step is governed by *Courant stability limit*. This limit can be interpreted that the time-step must be small enough to capture the wave traveling through an element which is equivalent to solving an oscillation at eigenfrequency of that element [37].

- At $\gamma = 0.5$, the method exhibits no additional damping rather than the natural one found in the system equations. The higher value of γ will demonstrate larger numerical damping in the system.

Commonly, the explicit integration provides higher accuracy of the solution than the implicit ones [52]. However, it requires the time-step Δt to be very small in order to preserve the level of stability and solution accuracy. There is no ideal suggestion for which method is superior to the other. It depends strongly to the application and the computing resources. In this work, the Newmark Implicit Integration method is employed while choosing $\beta = 0.25$ and $\gamma = 0.5$ as stated in the general guide-line above. With Newmark method, the simulation is allowed to be adjusted so that the size of time-step justify the computing resource while still provide the reasonable sampling rate of the simulation. However, time-step must be set to be small enough to capture the oscillation of an interested mode of the simulation. In this work, the simulation is appointed to serve as a visualization of the remote environment on the monitor. For a small continuous deformation to average human eyes, a dynamic simulation with sampling rate of $30\ Hz$ is sufficient which is equivalent to simulation time-step of $\approx 0.033\ s$.

4.5.2 Solution of linear algebraic equations

The last section discussed exclusively on Newmark Implicit Integration which formulates the linear dynamic governing system of equations 4.30 in a familiar set of linear algebraic equations. The displacement for the next simulation step is exhibited in equation 4.39 as vector $\boldsymbol{U}_{n+1}$. In order to solve the equation, equation 4.39

will be partitioned and rearranged after given boundary condition of each DoF and topology, also called *index*, as such

$$\begin{bmatrix} K^*_{\varphi\varphi} & K^*_{\varphi t} \\ K^*_{t\varphi} & K^*_{tt} \end{bmatrix}_n \cdot \begin{bmatrix} U_\varphi \\ U_t \end{bmatrix}_{n+1} = \begin{bmatrix} F_\varphi \\ F_t \end{bmatrix}^{ext}_{n+1} + \begin{bmatrix} F_\varphi \\ F_t \end{bmatrix}^{int}_n \quad (4.42)$$

with nodal displacements $U_{\varphi,n+1}$ are known from Dirichlet boundary conditions. $F^{ext}_{t,n+1}$ are given nodal external forces which are identical to Neumann boundary conditions. The solution of the rest nodal displacement in this linear algebraic equations for the next simulation step $n+1$ can then be written as

$$U_{t,n+1} = K^{*-1}_{tt,n} \left(F^{ext}_{t,n+1} + F^{int}_{t,n} - K^*_{t\varphi,n} \cdot U_{\varphi,n+1} \right) \quad (4.43)$$

After calculation of the nodal displacement U_{n+1}, its velocity and acceleration can be computed using equations 4.37 and 4.36 to obtain $\dot{U}_{n+1}$ and $\ddot{U}_{n+1}$ for the next simulation step. The computation of the internal nodal forces from the element circling the node F^{int}_n at current simulation step can be made with the complete registration of U_n, $\dot{U}_n$ and $\ddot{U}_n$. With earlier computation, the external nodal forces $F^{ext}_{\varphi,n+1}$ of the next simulation step can be computed as

$$F^{ext}_{\varphi,n+1} = K^*_{t\varphi,n} \cdot U_{\varphi,n+1} + K^*_{tt,n} \cdot U_{\varphi,n+1} - F^{int}_{\varphi,n} \quad (4.44)$$

Equation 4.44 is not necessary to be computed in this work since it provides the reaction force of corresponding to given nodal displacements of the same indexes. The nodal external forces obtained from the FEM contains part of the incision force. They only reflect the external force causing a dynamic deformation in the soft body. To obtain the actual incision force, an empirical force model is purposefully implemented which will be extensively discussed later in chapter 7.

4.6 Implementation

Primarily, a simulation problem must be formulated. Algorithm 1 demonstrates the initialization of a FEM simulation. The parameters, Young's modulus E, density ρ and Poisson ratio ν must be given. They are essential for determination of the material type of the considered solid body. The mesh generation is employed to spatially discrete the continuum domain of the considered soft body into finite number of tetrahedron elements at the current configuration $\mathcal{B}_\Gamma$ of the beginning simulation time $t_n = 0$. The topology is also obtained determining a successive index of nodal DoFs. The initial Neumann and Dirichlet boundary conditions must be given. Using the topology index, element mass, damping and stiffness matrices will be calculated and assembled to obtain the global system matrices. The formulation of the governing system of equations is done so that the Newmark implicit integration can be employed.

4.6 Implementation

Algorithm 1: Initialization of FEM simulation

Read material parameter:
E, ρ, ν
Simulation step indicator:
$t_n = 0$ and $n = 0$
Mesh generation: $\longleftarrow \mathcal{B}_\Gamma$ at $n = 0$
U, Initial topology, number of elements n_{ele}
Apply initial boundary conditions:
$U_{\varphi,0}, \dot{U}_{\varphi,0}, \ddot{U}_{\varphi,0}, F_{t,0}^{ext}$
Calculation initial system matrices:
M_0, D_0, K_0
Formulate dyn. governing system of equations: $\longrightarrow$ equation 4.38

Algorithm 2 demonstrates the pseudo code of a dynamic simulation procedure of the incision process employing FEM. In this work, the simulation is required to be constantly interacted with the external manipulation from a human operator via a haptic device. The variable r_C represents the scalpel position at current time t_n read from haptic device. If a collision between the scalpel and the soft body is detected and the rate of cut evolution is not zero, the material structural change is occurred to current configuration as such $\Xi_t : \Omega_\Gamma \to \mathcal{B}_\Gamma$. Thus, the remeshing is necessarily required to determine the change in current geometry of the simulation step.

The remeshing algorithm modifies U_n to exhibit the cut between nodes as discussed in section 4.4.2. Since the remeshing algorithm causes a successive change in the topology and the index of nodal DoFs denoted by *Current topology*, the assembly process must be called to update the dynamic governing system of equations. Before the boundary condition can be applied to the current dynamic governing system of equations, the remapping process must be done to ensure a physically correct transition between configurations of different topology. The updated dynamic governing system of equations can then be integrated to obtain the nodal displacements.

If a simulation cycle exhibits no collision, the simulation is carried out with the omission of update to the topology and system matrices. Therefore, the dynamic governing system of equations is solved to obtain the nodal displacements of the solid body at current geometry without remapping the boundary conditions.

The current simulation step is ended by update the graphical visualization with the calculated nodal displacements U_{n+1}.

Figure 4.5 illustrates the simulation dynamic geometric deformation of a solid-body using FEM with remeshing derived in this section. The initial number of degree-of-freedoms is 1089 with 1000 tetrahedron elements in total. As the cut evolves, the new nodes are added. Using the boundary of the elements surrounding the cut, the edge of the cut is directly presented.

4.6 Implementation

Algorithm 2: Incision simulation with FEM

for $Simulation = true$ do
  Read current scalpel position: $\longrightarrow r_{C,n}$
  if $r_{C,n}$ contacts with soft body and $\dot{r}_{C,n} \neq 0$ then
    Remeshing: $\Xi_t : \Omega_\Gamma \to \mathcal{B}_\Gamma$
    U_n, Current topology
    for Current topology do
      Remapping and assembly system matrices:
      $M_n(\mathcal{B}_\Gamma) \longleftarrow M_n(\Omega_\Gamma)$
      $D_n(\mathcal{B}_\Gamma) \longleftarrow D_n(\Omega_\Gamma)$
      $K_n(\mathcal{B}_\Gamma) \longleftarrow K_n(\Omega_\Gamma)$
      Remapping boundary conditions:
      $U_{\varphi,n}(\mathcal{B}_\Gamma) \longleftarrow U_{\varphi,n}(\Omega_\Gamma)$
      $\dot{U}_{\varphi,n}(\mathcal{B}_\Gamma) \longleftarrow \dot{U}_{\varphi,n}(\Omega_\Gamma)$
      $\ddot{U}_{\varphi,n}(\mathcal{B}_\Gamma) \longleftarrow \ddot{U}_{\varphi,n}(\Omega_\Gamma)$
      $F_{t,n}^{ext}(\mathcal{B}_\Gamma) \longleftarrow F_{t,n}^{ext}(\Omega_\Gamma)$
    end
  end
  else
    $M_n = M_{n-1}$
    $D_n = D_{n-1}$
    $K_n = K_{n-1}$
    Apply boundary conditions:
    $U_{\varphi,n}, \dot{U}_{\varphi,n}, \ddot{U}_{\varphi,n}, F_{t,n}^{ext}$
  end
  Solve dynamic governing system of equations: $\longrightarrow U_{n+1}, \dot{U}_{n+1}, \ddot{U}_{n+1}$
  Graphical visualization: $\longleftarrow U_{n+1}$
  Update simulation time: $n = n+1$, $t_{n+1} = t_n + \Delta t$
end

4.6 Implementation

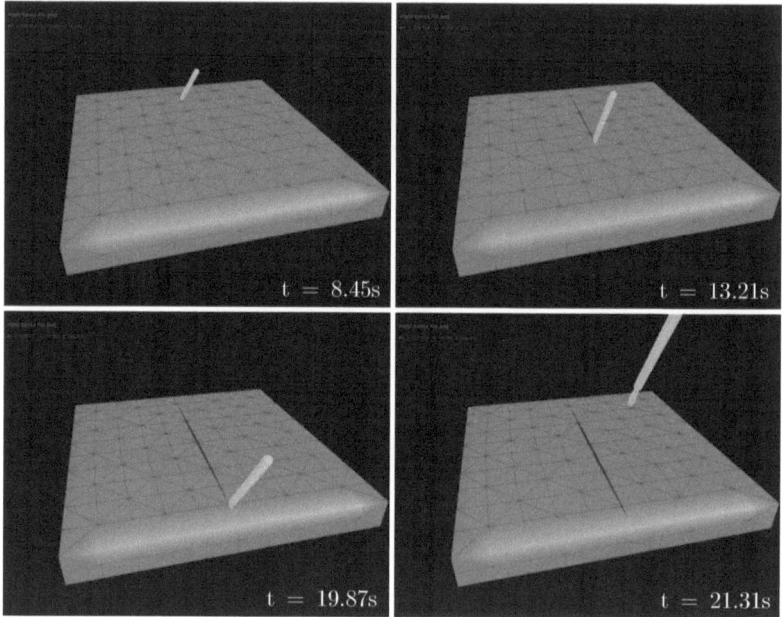

Figure 4.5: Incision simulation with geometry deformation using FEM with remeshing

5 Extended finite element approach

In this chapter, the XFEM is investigated as an alternative to the FEM with remeshing for a real-time incision simulation. The concept of XFEM was developed relying on the extrinsic enrichments and the local version of partition of unity but applies the enrichment terms only on a certain set of local subdomains. A brief graphical demonstration on the difference between conventional FEM and XFEM by treating a cut in a solid-body is depicted in Figure 5.1. On the left hand side is the spatial discretization depicted in section 4.1, Figure 4.1(b). FEM needs to discrete the solid-body so that the elements align and coincide the cut domain Γ. Whereas on the right hand side, Figure 5.1(b), is the XFEM letting the cut existing in the elements of the solid-body. The cut is instead differentiated by level-set method pointed by variable η and stays as part of the elements local properties but with additional local definitions.

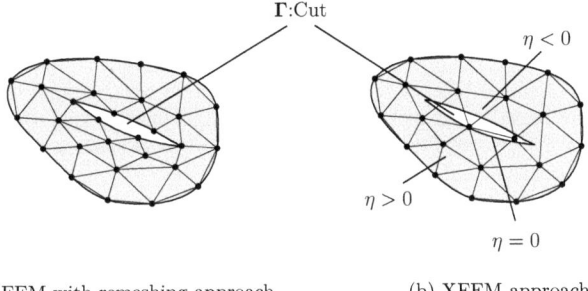

(a) FEM with remeshing approach   (b) XFEM approach

Figure 5.1: Illustration of a solid body with cut

Another benefit of using enrichment terms is that different discontinuities of various phenomena can be demonstrated together with minimal regard to the mesh and elements. The remeshing process of the conventional FEM may not be an option if only the domain preservation limiting the boundary and interface between materials using the boundary of the elements is employed in some problems such as non-isotropic and multi-phases material. Transfer of the boundary values can be confusing since the configuration mapping must be totally considered according to the remeshing. XFEM approach is developed to ease these complications specifically.

Since implementation of surgical simulator or modeling of incision simulation shares some of its features found in crack propagation. The approach is adopted gradually from the research community in the field of engineering, computer science and medicine. JEŘÁBKOVÁ exhibited an successful implementation of the surgical simulator based on XFEM in her dissertation. Interesting is also the work presented by VIGNERON which discusses the deformation in a soft body as the result of both surgical cutting and tearing [96]. His works shows the ability of XFEM to consider two discontinuities at the same time. This chapter is devoted to the discussion on treating the incision or cut as a discontinuity in a solid material which is the background principle of the XFEM approach. The approach will take the kinematic of the incision developed earlier in chapter 3 and derived the incision as part of the displacement discretization in element directly without considering the remeshing process. Since XFEM is actually the extended or variation of the conventional FEM discussed elaborately in chapter 4, some derivations and assumptions will be adopted in this section with notices. The derivation and some illustrations regarding XFEM used in this chapter can also be reviewed from [16, 39, 46, 58, 63] if a more extensive explanation on the XFEM approach is required.

5.1 Displacement approximation with discontinuities

XFEM can be assumed to be a classical FEM capable of handling arbitrary strong and weak discontinuities [58]. The concept of XFEM is to enrich the approximation spaces so that it is capable of reproducing certain features of the problem of interest. The enrichments are applied only in a certain local sub-domain in order to increase the order of completeness of the solution. In other words, it improves the accuracy of the approximation by including the information obtained from the analytical solution in the local element domain directly. Employing XFEM in a simulation requires initially a conventional mesh similar to one from FEM. Then, during the simulation, the location of discontinuities are considered. Accordingly, new degrees of freedom are added to the conventional governing system of equations in the selected nodes near the discontinuities.

Let consider the nodal reference configuration in a solid body $\boldsymbol{X}$, also assumed is that a discontinuity is exhibited in an element of this solid body which has n nodes. The following displacement approximation $\boldsymbol{u}^h$ is proposed by BELYTSCHKO AND BLACK for crack propagation simulation [14]. The formulation of the discrete kinematics shall not be repeated but can be reviewed from section 4.1.

$$\boldsymbol{u}^h = \boldsymbol{u}^{std} + \boldsymbol{u}^{enr} = \sum_{j=1}^{n} N_j(\boldsymbol{X})\boldsymbol{u}_j + \sum_{k=1}^{m} N_k(\boldsymbol{X})\boldsymbol{\psi}(\boldsymbol{X})\boldsymbol{a}_k \tag{5.1}$$

The first term of equation 5.1, $\boldsymbol{u}^{std}$ is recognized as the standard displacement approximation seen in equation 4.7 of FEM. Where the second term, $\boldsymbol{u}^{enr}$ refers to the additional approximation or enrichment term due to discontinuity in the elements containing cut. $\boldsymbol{u}_j$ is the nodal displacement vector of the regular degrees of freedom found in the FEM. $\boldsymbol{a}_k$ denotes the added set of DoFs m to the standard FEM. $\boldsymbol{\psi}(\boldsymbol{X})$ is the discontinuous enrichment function defined in the set of nodes near a discontinuity.

The enrichment function $\psi(\boldsymbol{X})$, in general, can be chosen by applying an appropriate analytical solution according to the type of discontinuity. It is the fact that different types of phenomena require specific mathematic explanations. The shape function of the enrichment term N_k is not required to be identical to that of the standard FEM N_j. It depends on the physical type of phenomena. In this work, it is assumed that $N_j = N_k$ and are identical to the shape functions for a tetrahedral element.

With the concept of XFEM, different and multiple types of discontinuity are allowed to be expressed in the same displacement approximation by adding just additional terms coincide with the number and type of discontinuities as

$$\boldsymbol{u}^h = \sum_{j=1}^{n} N_j(\boldsymbol{X})\boldsymbol{u}_j + \sum_{l=1}^{np}\sum_{k=1}^{m} N_k(\boldsymbol{X})\psi^l(\boldsymbol{X})\boldsymbol{a}_k^l \tag{5.2}$$

with np denotes the number of discontinuities existing in the element. The original mesh generated for the standard displacement approximation is mapped to the vector $\boldsymbol{u}$ while the discontinuities are performed by the enrichment terms. The XFEM does not require the remeshing but instead employ an enrichment function for each discontinuities. This is an advantage if the simulation involves in multiple cut in the same element [46] or in an complicate interface problems such as the simulation of multi-phase flow where the material properties are changed gradually as they are mixed together [21,35].

5.2 Modeling of strong discontinuous fields

Incision, cut or crack are all considered to be one of strong discontinuities in an element since it completely break the material bonds on both side of a cut plane. As mentioned earlier that modeling of a discontinuity in XFEM requires a proper enrichment function suitable for a specific discontinuity.

To demonstrate the effect of enrichment functions, Figure 5.2 is brought to consider an one-dimensional deformation problem. The problem consists of three elements and four nodes. N_j with $j = 1, 2, 3, 4$ are the shape functions satisfy the Kronecker Delta property and the Partition of Unity discussed in section 4.2. A strong discontinuity assumed to be a cut exists arbitrarily at $X_c(\eta = 0)$ in the second element. The second element is therefore divided or separated into two bodies which is quantitatively described as one staying above the cut $(\eta > 0)$ and one staying below the cut $(\eta < 0)$.

XFEM approach requires only the second element to be locally enriched. The enrichment affects accordingly second and third nodes, whereas first and forth nodes are not influenced by the cut and the enrichment does not required in these domains. The essential question is how to mathematically demonstrate the cut in the element by choosing an enrichment function $\psi(\boldsymbol{X})$ in equation 5.1. For a strong discontinuity such as the given example, two enrichment functions are often employed which are the *Heaviside* and the *Shifted* functions.

5.2 Modeling of strong discontinuous fields

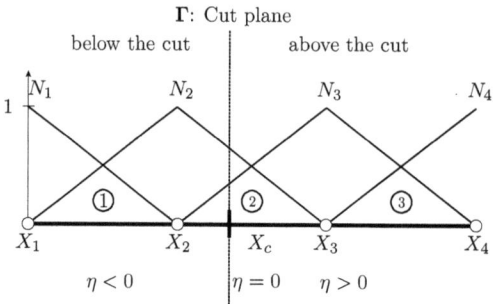

Figure 5.2: A cut plane in a one-dimensional problem

5.2.1 Heaviside function

The Heaviside function is the simplest and widely used enrichment function which is also easily recognized as sgn() function as shown in equation 5.3 [14].

$$\psi(\boldsymbol{X}) = H(\boldsymbol{\eta}) = \begin{cases} +1 & \forall \eta > 0 : \text{above the cut plane} \\ -1 & \forall \eta < 0 : \text{below the cut plane} \end{cases} \quad (5.3)$$

Figure 5.3(a) demonstrates the Heaviside function enrichment results in a discontinuity in shape functions N_2 and N_3 which are used for displacement approximation of in the second element. The displacement approximation affected by the discontinuities modeled by Heaviside function is therefore known as

$$\boldsymbol{u}^h(\boldsymbol{X}_i) = \boldsymbol{u}_i + H(\boldsymbol{\eta}_i)\boldsymbol{a}_i \neq \boldsymbol{u}_i \quad (5.4)$$

One problem using Heaviside as enrichment function is the enrichment term will not satisfy the Kronecker delta property [46,51]. Figure 5.3(b) explains this problem that the approximations from XFEM with Heaviside on the first and third elements are no longer exactly identical to the approximation provided by the conventional ones. The value of the field variable $\boldsymbol{u}(\boldsymbol{X})$ on the enriched node i is not equal to the actual nodal displacement value $\boldsymbol{u}_i$. Thus, employing the Heaviside function produces unwanted discontinuities affecting the displacement in the neighbor elements of the enriched element. $\boldsymbol{u}^h$ obtained in equation 5.4 does not cater a proper physical meaning of the actual nodal displacement and need to be used with additional modifications.

5.2.2 Shifted function

A solution preventing the shortcoming occurred from Heaviside function is called Shifted enrichment function shown in equation 5.5 [31]. In Figure 5.4(a), the effect

5.2 Modeling of strong discontinuous fields

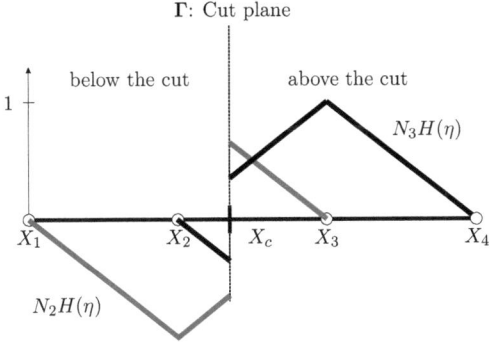

(a) Discontinuities from Heaviside function in second element

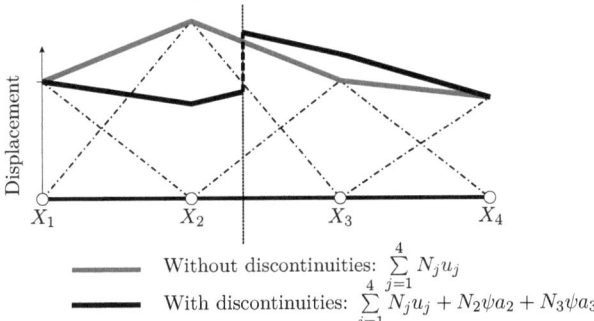

(b) Total displacement with discontinuities by Heaviside function

Figure 5.3: Enrichment with Heaviside function

of Shifted enrichment function on the shape functions is demonstrated.

$$\psi(\boldsymbol{X}) = \frac{1}{2}\left(H(\boldsymbol{\eta}) - H_i\right) \tag{5.5}$$

H_i is the value of $H(\boldsymbol{\eta})$ on the i-th node. With $1/2$ factor, the interpolation is ensured to satisfy the Kronecker delta property. Figure 5.4(b) points that the discontinuity using Shifted function happens only on the second element alone. The displacement of the neighbor elements without discontinuity stay unaffected in this case. Hence the physical interpretation of the nodal displacement is correctly preserved. The approximation of the nodal displacement using a Shifted enrichment function is then found as

$$\boldsymbol{u}^h(\boldsymbol{X}_i) = \boldsymbol{u}_i + \frac{1}{2}\left(H(\boldsymbol{\eta}_i) - H_i\right)\boldsymbol{a}_i = \boldsymbol{u}_i \tag{5.6}$$

5.2 Modeling of strong discontinuous fields

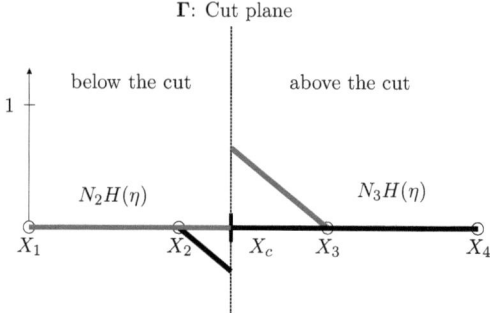

(a) Discontinuities from Shifted function in second element

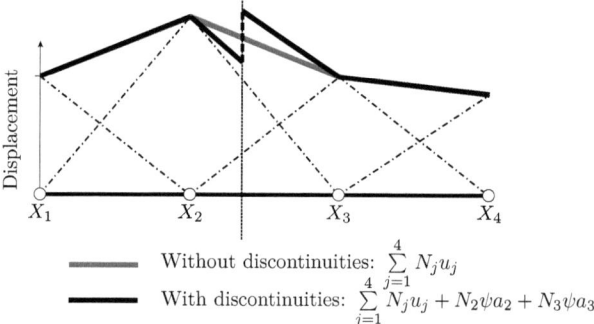

(b) Total displacement with discontinuities by Shifted function

Figure 5.4: Enrichment with Shifted function

In this work, the demonstration and simulation of a strong discontinuity from cut with a sharp scalpel is developed. The benefits of Shifted enrichment function are obvious that the implementation is not complicated and the physical interpretation of the cut is provided, thus the additional modification of the obtained approximation can be used as the actual displacement for the simulation.

The enrichment functions for each different discontinuities can be promptly put into an element to demonstrate multiple phenomenas. Other than the Shifted enrichment function presenting strong discontinuity, the other discontinuities e.g. cut evolving due to tearing because of the external tension force can be approximate together using another proper enrichment function similar to work of VIGNERON [96]. Also, the work of MOHAMMADI is recommended which discusses on various enrichment examples [58].

5.3 Discrete governing system of equations with XFEM

Section 5.3 discusses on how to add the additional analytic solution to the displacement approximation and improve its completeness and accuracy. This analytic approximation comes in form of the discontinuity enrichment function. For incision simulation, where the cut in a solid-body is treated by XFEM as a type of strong discontinuities, the Shifted enrichment approach is seen as a suitable solution and is adopted throughout the upcoming implementation in this work.

In order to obtain a discrete governing system of equations necessary for the explanation of the dynamic behavior of a solid-body, similar to the approach taken in section 4.2, the discrete kinematics with discontinuity derived in previous sections will be considered while the kinematic of incision known from section 3 will be recalled again for XFEM derivation.

To begin with, the displacement approximation from equation 5.1 u^h is rewritten in matrix format as

$$u^h = \begin{bmatrix} N & \psi N \end{bmatrix} \cdot \begin{bmatrix} u^{std} \\ a \end{bmatrix} \tag{5.7}$$

Vector u^h puts together the standard approximation of the FEM u^{std} and adds the additional DoFs a from the discontinuity approximation. In a similar manner, the strain-displacement matrix B is extended to support these additional DoFs as

$$B^h = \begin{bmatrix} B & \psi B \end{bmatrix} \tag{5.8}$$

Considering Figure 5.5 gives a close perspective of the physical interpretation explaining how XFEM approach works. In the Figure, a tetrahedral element with total volume V is cut arbitrarily and therefore divided into two volumes. V_a and V_b are referred to *volume above the cut plane* and *volume below the cut plane* respectively. Thus, to obtain an enriched element stiffness matrix $k^{h(e)}$, the integral for the element stiffness matrix over total volume can accordingly be done separately [16, 46]

$$k^{h(e)} = \int_{\Omega_e} \left(B^h\right)^T \hat{C} B^h \mathrm{d}V = \int_{V_a} \left(B^h\right)^T \hat{C} B^h \mathrm{d}V + \int_{V_b} \left(B^h\right)^T \hat{C} B^h \mathrm{d}V \tag{5.9}$$

Substituting B^h from equation 5.8 and obtain

$$k^{h(e)} = \int_{\Omega_e} \begin{bmatrix} B^T \hat{C} B & B^T \hat{C} \psi B \\ \psi B^T \hat{C} B & \psi B^T \hat{C} \psi B \end{bmatrix} \mathrm{d}V = \begin{bmatrix} k^{uu(e)} & k^{ua(e)} \\ k^{au(e)} & k^{aa(e)} \end{bmatrix} \tag{5.10}$$

5.3 Discrete governing system of equations with XFEM

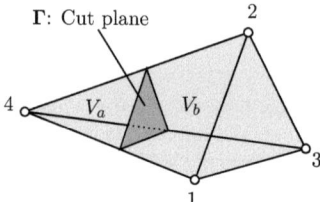

Figure 5.5: Dissected tetrahedral element: cut plane divided the element into two volumes and defining V_a and V_b for above and below the cut respectively

Accordingly, the stiffness matrix is divided into four parts for a simple interpretation. $k^{uu(e)}$ coincides with the standard stiffness matrix and is identical to the one introduced in equation 4.24. The other three correspond to the additional DoFs a. In case of linear Finite Element, the strain-displacement matrix B is constant. This property is legally conveyed to XFEM approximation and used here as well. Therefore,

$$k_{ij}^{uu(e)} = k_{ij}^{(e)} \tag{5.11}$$

$$k_{ij}^{ua(e)} = \left(\frac{V_a}{V}\psi_{aj} + \frac{V_b}{V}\psi_{bj}\right) k_{ij}^{(e)} \tag{5.12}$$

$$k_{ij}^{au(e)} = \left(\frac{V_a}{V}\psi_{ai} + \frac{V_b}{V}\psi_{bj}\right) k_{ij}^{(e)} \tag{5.13}$$

$$k_{ij}^{aa(e)} = \left(\frac{V_a}{V}\psi_{ai}\psi_{aj} + \frac{V_b}{V}\psi_{bi}\psi_{bj}\right) k_{ij}^{(e)} \tag{5.14}$$

For Shifted enrichment function, the enriched stiffness matrix can be calculated as [46]

$$k_{ij}^{uu(e)} = k_{ij}^{(e)} \tag{5.15}$$

$$k_{ij}^{ua(e)} = \begin{cases} \dfrac{V_b}{V}k_{ij}^{(e)}; & \text{for } H_j = -1 \\[6pt] \dfrac{-V_a}{V}k_{ij}^{(e)}; & \text{for } H_j = +1 \end{cases} \tag{5.16}$$

$$k_{ij}^{au(e)} = \begin{cases} \dfrac{V_b}{V}k_{ij}^{(e)}; & \text{for } H_i = -1 \\[6pt] \dfrac{-V_a}{V}k_{ij}^{(e)}; & \text{for } H_i = +1 \end{cases} \tag{5.17}$$

5.3 Discrete governing system of equations with XFEM

$$k_{ij}^{aa(e)} = \begin{cases} \dfrac{V_b}{V}\mathbf{k}_{ij}^{(e)}; & \text{for } H_i = H_j = -1 \\ \dfrac{V_a}{V}\mathbf{k}_{ij}^{(e)}; & \text{for } H_i = H_j = +1 \\ 0; & \text{for } H_i \neq H_j \end{cases} \quad (5.18)$$

As discussed in section 3.5, the incision promotes no total mass lost. Therefore, the expression of the enriched element mass matrix $\mathbf{m}^{h(e)}$ is similar to that of the enriched stiffness matrix. The enriched shape function from equation 5.7 is applied to the element mass matrix from equation 4.25 to obtain

$$\begin{aligned}\mathbf{m}^{h(e)} &= \int_{\Omega_e} \begin{bmatrix} \mathbf{N} & \psi\mathbf{N} \end{bmatrix}^T \rho \begin{bmatrix} \mathbf{N} & \psi\mathbf{N} \end{bmatrix} \mathrm{d}V \\ &= \int_{\Omega_e} \rho \begin{bmatrix} \mathbf{N}\mathbf{N} & \mathbf{N}\psi\mathbf{N} \\ \psi\mathbf{N}\mathbf{N} & \psi\mathbf{N}\psi\mathbf{N} \end{bmatrix} \mathrm{d}V \\ &= \int_{V_a} \begin{bmatrix} \mathbf{N} & \psi\mathbf{N} \end{bmatrix}^T \rho \begin{bmatrix} \mathbf{N} & \psi\mathbf{N} \end{bmatrix} \mathrm{d}V + \int_{V_b} \begin{bmatrix} \mathbf{N} & \psi\mathbf{N} \end{bmatrix}^T \rho \begin{bmatrix} \mathbf{N} & \psi\mathbf{N} \end{bmatrix} \mathrm{d}V \\ &= \begin{bmatrix} \mathbf{m}^{uu(e)} & \mathbf{m}^{ua(e)} \\ \mathbf{m}^{au(e)} & \mathbf{m}^{aa(e)} \end{bmatrix} \end{aligned} \quad (5.19)$$

For a linear Finite Element approximation, the shape functions are kept constant and allow the enriched element mass matrix to be integrated

$$m_{ij}^{uu(e)} = m_{ij}^{(e)} \quad (5.20)$$

$$m_{ij}^{ua(e)} = \begin{cases} \dfrac{V_b}{V}\mathbf{m}_{ij}^{(e)}; & \text{for } H_j = -1 \\ \dfrac{-V_a}{V}\mathbf{m}_{ij}^{(e)}; & \text{for } H_j = +1 \end{cases} \quad (5.21)$$

$$m_{ij}^{au(e)} = \begin{cases} \dfrac{V_b}{V}\mathbf{m}_{ij}^{(e)}; & \text{for } H_i = -1 \\ \dfrac{-V_a}{V}\mathbf{m}_{ij}^{(e)}; & \text{for } H_i = +1 \end{cases} \quad (5.22)$$

$$m_{ij}^{aa(e)} = \begin{cases} \dfrac{V_b}{V}\mathbf{m}_{ij}^{(e)}; & \text{for } H_i = H_j = -1 \\ \dfrac{V_a}{V}\mathbf{m}_{ij}^{(e)}; & \text{for } H_i = H_j = +1 \\ 0; & \text{for } H_i \neq H_j \end{cases} \quad (5.23)$$

5.3 Discrete governing system of equations with XFEM

Analogously, the element damping matrix can be calculated as of the FEM approach using Rayleigh formulation. The derivation of damping is explicitly written as follow

$$d_{ij}^{uu(e)} = a\, m_{ij}^{(e)} + b\, k_{ij}^{(e)} \tag{5.24}$$

$$d_{ij}^{ua(e)} = \begin{cases} \dfrac{V_b}{V} d_{ij}^{(e)}; & \text{for } H_j = -1 \\ \dfrac{-V_a}{V} d_{ij}^{(e)}; & \text{for } H_j = +1 \end{cases} \tag{5.25}$$

$$d_{ij}^{au(e)} = \begin{cases} \dfrac{V_b}{V} d_{ij}^{(e)}; & \text{for } H_i = -1 \\ \dfrac{-V_a}{V} d_{ij}^{(e)}; & \text{for } H_i = +1 \end{cases} \tag{5.26}$$

$$d_{ij}^{aa(e)} = \begin{cases} \dfrac{V_b}{V} d_{ij}^{(e)}; & \text{for } H_i = H_j = -1 \\ \dfrac{V_a}{V} d_{ij}^{(e)}; & \text{for } H_i = H_j = +1 \\ 0; & \text{for } H_i \neq H_j \end{cases} \tag{5.27}$$

The element nodal external forces are comprised with the discrete body force and the discrete surface traction force, see section 4.2 equations 4.18 and 4.19. To support the additional DoFs a, the element nodal external forces are extended. Here arises a discussion on how extended forces f^h should be determined. Doing so, the derivation of an element nodal external forces is adopted while the standard shape function is substituted with $\begin{bmatrix} N & \psi N \end{bmatrix}$. This yields

$$f^h = \begin{bmatrix} f^u & f^a \end{bmatrix}^T = \int_{\Omega_e} \begin{bmatrix} N & \psi N \end{bmatrix}^T \rho \mathbf{b} \, dV + \int_{\partial\Omega_e} \begin{bmatrix} N & \psi N \end{bmatrix}^T \tilde{T}_0 \, dA \tag{5.28}$$

This approximation can be separately integrated on two volumes and surfaces using the earlier assumption. With constant shape functions for linear Finite Element approximation, the extended nodal external force is found as

$$f_i^{u(e)} = f_i^{ext(e)} = m_{ij}^{uu(e)} b_i + \int_{\Omega_e} N_i \tilde{T}_{i_0} \, dA \tag{5.29}$$

$$f_i^{a(e)} = \begin{cases} \dfrac{V_b}{V} m_{ij}^{(e)} b_i + \int_{A_b} N_i \tilde{T}_{i_0} \, dA; & \text{for } H_j = -1 \\ \dfrac{-V_a}{V} m_{ij}^{(e)} b_i - \int_{A_a} N_i \tilde{T}_{i_0} \, dA; & \text{for } H_j = +1 \\ 0; & \text{for } H_j \neq H_i \end{cases} \tag{5.30}$$

In case of multiple incisions in an element, the displacement approximation derived in equation 5.2 can be used. This approach allows multiple discontinuities to coexist in an element using repeating enrichment terms.

5.4 Enrichment nodes selection and assembly

The discrete kinematics in the last section are derived for an element subjected to a cut after XFEM. The dissected elements and their nodes must be selected for the local enrichment treatment. It results in the dimension expansion of the system matrices due to the additional DoFs. To obtain the governing system of equations after the cut, the system matrices must be assembled before the integrator can solve for the deformation displacement.

5.4.1 Enrichment nodes selection

Figure 5.6 depicts a discrete soft body containing a cut. The nodes of all elements containing cut selected for the enrichments and marked with circles. The other elements and their nodes are left unmodified.

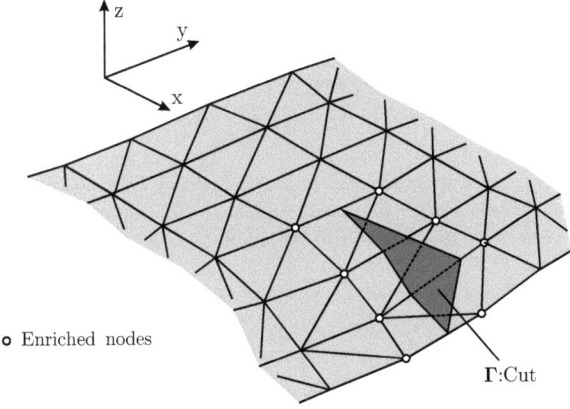

Figure 5.6: Selection of elements and nodes for enrichment

Similarly to the remeshing method required by conventional FEM approach discussed, the evolution of the cut must be updated at each simulation step. In this work, the position of the scalpel is sampled accordingly. The incision part is determined by linear interpolation between current and last position. It is used to determine the divided volumes of the dissected elements.

5.4 Enrichment nodes selection and assembly

After nodes selection, the discontinuities are added to each shape functions of the dissected elements, while the calculation of extended terms in element stiffness, mass and damping matrices are carried out. The additional DoFs are put into and extend the displacement vector accordingly. Since the added DoFs are put purposefully to increase the approximation accuracy, they coincide with the number of shape functions in an element required to be enhanced. If a tetrahedral element is thoroughly cut, all four shape functions are required to be enriched. Analogously, 12 additional DoFs are necessary to enrich this element. For example, an element stiffness matrix is extended after a cut as

$$\boldsymbol{k}^{h(e)} = \left[\begin{array}{ccc|ccc} k^{uu}_{1,1} & \cdots & k^{uu}_{1,12} & k^{ua}_{1,1} & \cdots & k^{ua}_{1,12} \\ \vdots & \ddots & \vdots & \vdots & \ddots & \vdots \\ k^{uu}_{12,1} & \cdots & k^{uu}_{12,12} & k^{ua}_{12,1} & \cdots & k^{ua}_{12,12} \\ \hline k^{au}_{1,1} & \cdots & k^{au}_{1,12} & k^{aa}_{1,1} & \cdots & k^{aa}_{1,12} \\ \vdots & \ddots & \vdots & \vdots & \ddots & \vdots \\ k^{au}_{12,1} & \cdots & k^{au}_{12,12} & k^{aa}_{12,1} & \cdots & k^{aa}_{12,12} \end{array}\right] \in \mathbb{R}^{24 \times 24} \quad (5.31)$$

The same computation method can be applied to mass and damping matrices for each tetrahedral element containing cut.

5.4.2 Assembly of enrichment DoFs

The added DoFs $\boldsymbol{a}$ as the consequence from the enriched nodes are staying as the extended approximation shown in equation 5.1 and 5.2. XFEM provides a systematic expansion of the governing system of equations which requires only the $\boldsymbol{a}$ to stay right after the standard DoFs $\boldsymbol{u}$. The assembly can be implemented to concentrate straightforwardly on these expanding regions.

The expansion in system matrices to support the additional DoFs is demonstrated by the global stiffness matrix $\boldsymbol{K}^h$ in equation 5.32. The dimension of the global stiffness matrix is extended dependently to the number of the dissected elements n_{cut}. Since 12 additional DoFs $\boldsymbol{a}$ are required for each completely dissected element, the dimension of the global stiffness matrix can be calculated as $\mathbb{R}^{(3n_{node}+12n_{cut}) \times (3n_{node}+12n_{cut})}$.

$$\boldsymbol{K}^h = \left[\begin{array}{c|c} \bigcup\limits_{e=1}^{n_{ele}} \boldsymbol{k}^{uu(e)} & \bigcup\limits_{k=1}^{n_{cut}} \boldsymbol{k}^{ua(e)} \\ \hline \bigcup\limits_{k=1}^{n_{cut}} \boldsymbol{k}^{au(e)} & \bigcup\limits_{k=1}^{n_{cut}} \boldsymbol{k}^{aa(e)} \end{array}\right] \in \mathbb{R}^{(3n_{node}+12n_{cut}) \times (3n_{node}+12n_{cut})} \quad (5.32)$$

The assembly process of the expanded stiffness matrix can be done by separate the standard part to the extended parts. This yields great advantage as the standard global stiffness matrix $\boldsymbol{K}^{uu} = \bigcup\limits_{e=1}^{n_{ele}} = \boldsymbol{k}^{uu(e)}$ is computed only once during problem initialization. It is preserved and remain unchanged for linear XFEM. The extended parts are systematically and separately assembled before combined to $\boldsymbol{K}^{uu}$ to form a total enriched global stiffness matrix K^h. This procedure is also valid for enriched global mass and damping matrices.

5.5 Dynamic simulation

After assembly process, the extended dynamic governing system of equations is obtained as presented in equation 5.33.

$$F^h = M^h \ddot{u}^h + D^h \dot{u}^h + K u^h \tag{5.33}$$

or

$$\begin{bmatrix} F^{ext} \\ F^a \end{bmatrix} = \begin{bmatrix} M^{uu} & M^{ua} \\ M^{au} & M^{aa} \end{bmatrix} \cdot \begin{bmatrix} \ddot{U} \\ \ddot{A} \end{bmatrix} + \begin{bmatrix} D^{uu} & D^{ua} \\ D^{au} & D^{aa} \end{bmatrix} \cdot \begin{bmatrix} \dot{U} \\ \dot{A} \end{bmatrix} + \begin{bmatrix} K^{uu} & K^{ua} \\ K^{au} & K^{aa} \end{bmatrix} \cdot \begin{bmatrix} U \\ A \end{bmatrix}$$

In order to solve for the solution, a proper integration is required. XFEM encounters no restriction to employ the Newmark Implicit Integration discussed in section 4.5.1. The integration on XFEM is rather straight forward. The Newmark Implicit Integration parameters can also adopted without any necessity of modification from its conventional setting from FEM in order to remain unconditionally stable.

Since XFEM approach does not alter the topology of the standard FEM part, the initial boundary conditions applied earlier during initialization does not require an update to be mapped with the current topology. Similarly, the displacements u obtained from the integration can be mapped directly exhibits the actual geometry affected to discontinuities without.

Another benefit of using XFEM is that the governing system of equations are reversible to the origin state or any state back in time. This is viable due to the preservation of the standard FEM part and systematic expansion of the governing system of equations. This can be an advantage if the evolution of the cut or a discontinuity is history dependent.

5.6 Implementation

The implementation of the XFEM is similar to algorithm 2 of FEM. The simulation procedures begin with the initialization or formulation of the simulation problem. In this regard, the XFEM can be initialized using a similar initialization algorithm to that of FEM. The material properties, Young's modulus E, density ρ and Poisson ratio ν must be provided.

The Lagrangian mesh generates the mesh topology which also determines the index of displacement vector U and the number of elements found in the solid body. The initial Neumann and Dirichlet boundary conditions must be applied. The initial global mass M_0^h, damping D_0^h and K_0^h stiffness matrices can now be calculated.

5.6 Implementation

Without enrichment at the beginning of the simulation, these system matrices are identical to $\boldsymbol{M}^{uu}, \boldsymbol{D}^{uu}$ and $\boldsymbol{K}^{uu}$.

The formulation of the dynamic system of governing equations is done so that the equation is ready to be solved using Newmark implicit integration method discussed in section 4.5.1.

It is worth to note that the initializations of FEM and XFEM can be shared due to their similarity. XFEM can be considered a variation of the FEM simulation. Thus, if the simulations with both approaches are required, the implementation in a software can be done together without additional effort.

Algorithm 3: Initialization of XFEM simulation

Read material parameter:
E, ρ, ν
Simulation step indicator:
$t_n = 0$ and $n = 0$
Mesh generation: $\longleftarrow$ $\mathcal{B}_\Gamma$ at $n = 0$
U, $topology$, $number\ of\ elements\ n_{ele}$
Apply initial boundary conditions:
$\boldsymbol{U}_{\varphi,0}$, $\dot{\boldsymbol{U}}_{\varphi,0}$, $\ddot{\boldsymbol{U}}_{\varphi,0}$, $F_{t,0}^{ext}$
Calculation initial system matrix:
$\boldsymbol{M}_0^h = \boldsymbol{M}^{uu}$, $\boldsymbol{D}_0^h = \boldsymbol{D}^{uu}$, $\boldsymbol{K}_0^h = \boldsymbol{K}^{uu}$
Formulate dyn. governing system of equation: $\longrightarrow$ equation 5.33

Algorithm 4 is the pseudo code of the incision simulation procedure using XFEM. During the simulation, the scalpel position $\boldsymbol{r}_{C,n}$ is read constantly at each simulation step. If a collision between scalpel and the soft body is detected and the rate of cut evolution $\dot{\boldsymbol{r}}_{C,n}$ is not zero, the structural change $\boldsymbol{\Xi}_t : \Omega_\Gamma \to \mathcal{B}_\Gamma$ will be determined. The nodes of the dissected elements during the last simulation step must be selected and enriched. In this work, the elements are only allowed to be cut thoroughly. Thus, the number of additional DoFs $\boldsymbol{a}$ depends on the number of dissected elements as discussed in section 5.4.1.

The displacement vector $\boldsymbol{U}_n^h$ and the enriched force vector $\boldsymbol{F}_n^h$ are extended adding the $\boldsymbol{a}$ after its highest indexes. Accordingly, the enriched element mass, damping, stiffness and nodal forces of each dissected elements are computed and assembled to update the global matrices $\boldsymbol{M}_n^h$, $\boldsymbol{D}_n^h$, $\boldsymbol{K}_n^h$ and $\boldsymbol{F}_n^h$ to the current simulation step.

In case, the collision between the scalpel tip and the soft body does not occur or the $\dot{\boldsymbol{r}}_{C,n}$ is zero, it is implied that the structural change $\boldsymbol{\Xi}_t = 0$ at the current simulation step. Therefore, the system matrices and nodal external forces of the last simulation step are adopted as the current system matrices since they are identical.

5.6 Implementation

Applying the boundary conditions is a piece-wised mapping to the original topology of the standard DoFs as of FEM approach. The additional DoFs a are not given but to be solved with the unknown nodal displacements. Solving the dynamic governing system of equations will provide the displacement and its derivative for the next simulation step $n+1$ which is used accordingly to update the graphical visualization.

Algorithm 4: Incision simulation with XFEM

---

for *Simulation = true* do
    Read current scalpel position: $\longrightarrow r_{C,n}$
    if $r_{C,n}$ *contacts with soft body* and $\dot{r}_{C,n} \neq 0$ then
        Determination of dissected elements: $\Xi_t : \Omega_\Gamma \to \mathcal{B}_\Gamma$
        Element number for enrichment : e
        Enrichment of corresponding nodes:
        $\longrightarrow U_n^h = \begin{bmatrix} U^{std} & a \end{bmatrix}_n^T$, $F_n^h = \begin{bmatrix} F^{ext} & F^a \end{bmatrix}_n^T$
        for *all enrichment node* := e do
            Calculation and assembly enriched system matrices:
            $M_n^h \longleftarrow m_n^{ua(e)}, m_n^{au(e)}, m_n^{au(e)}$
            $D_n^h \longleftarrow d_n^{ua(e)}, d_n^{au(e)}, d_n^{au(e)}$
            $K_n^h \longleftarrow k_n^{ua(e)}, k_n^{au(e)}, k_n^{au(e)}$
            $F_n^h \longleftarrow f_n^{a(e)}$
        end
    end
    else
        $M_n^h = M_{n-1}^h$
        $D_n^h = D_{n-1}^h$
        $K_n^h = K_{n-1}^h$
        $F_n^h = F_{n-1}^h$
    end
    Apply boundary conditions:
    $U_{\varphi,n}^{std}, \dot{U}_{\varphi,n}^{std}, \ddot{U}_{\varphi,n}^{std}, F_{t,n}^{ext}$
    Solve dynamic governing system of equations: $\longrightarrow U_{n+1}^{std}, \dot{U}_{n+1}^{std}, \ddot{U}_{n+1}^{std}$
    Graphical visualization: $\longleftarrow U_{n+1}^{std}$
    Update simulation time: $n = n+1$, $t_{n+1} = t_n + \Delta t$
end

Figure 5.7 illustrates the simulation of dynamic geometry deformation of a solid-body using XFEM. The number of degree-of-freedoms at the initial state is 1,188 and 1,100 tetrahedron elements. The evolution of the cut is demonstrated directly in the elements. In contrast to FEM with remeshing which the boundary of the elements surrounding the cut representing the edge, XFEM does require additional

5.6 Implementation

post-processing of the visualization of the cut by mapping the enrichment degree-of-freedoms as the edge of the cut instead.

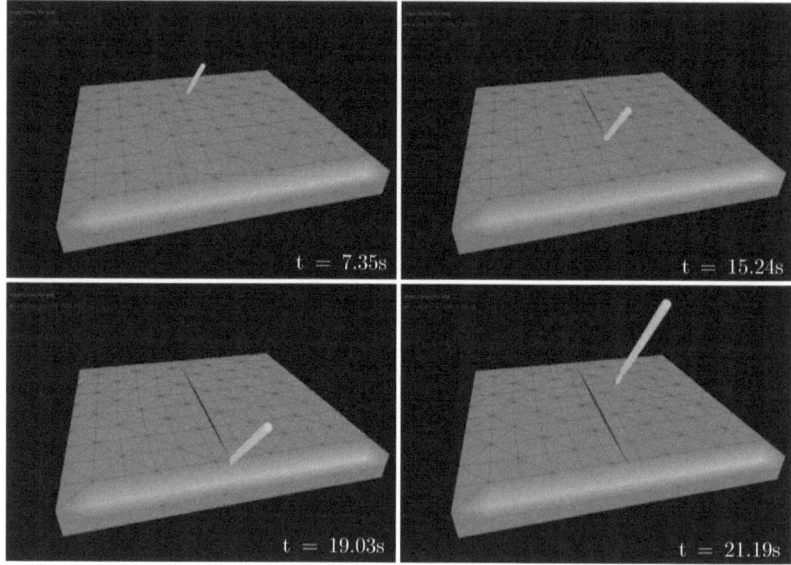

Figure 5.7: Incision simulation with geometry deformation using XFEM

6 Remarks on implementation practicality

A practical modeling of a soft body in fast transient applications such as the incision process is a major challenge of this work. It is because the simulation is aimed to be used as a contact model for telepresence application. Obviously, the real-time visualization capability must be obtained. Two approaches are proposed, FEM and XFEM. This chapter will discuss on their practicalities as a developing tool of a contact model.

6.1 Finite element method with remeshing

FEM is appointed to demonstrate the dissection of elements when a cut is exhibited. Doing so the Lagrange mesh which is fixed to specific material points must be reconsidered. An example remeshing algorithm from section 4.4.1 is employed for this task to handle a mesh condition which represents the actual geometry of the soft body at each simulation step. The remeshing algorithm in this work is developed to constrain the incision on a single vertical plane and only between the nodes. As a result, the topology of the governing system of equations is changed and the dimension of system matrices is expanded to associate the increasing number of DoFs from the additional nodes surrounding the cut. The dynamic governing system of equations is required to be reassembled similarly to the initializing of the simulation. Applying the boundary conditions are subjected to rearranged topology. Thus, the remapping before applying the boundary conditions from the predecessor to the current configuration must be made.

It must be noted that, for a general case, the implementation efficiency of FEM for an incision simulation depends strongly on the corporate remeshing algorithm. If the incision was made in the middle of the elements, the number of added elements may not be zero since remeshing algorithm must substitute the dissected tetrahedron elements with generally smaller ones [19,49]. Thus, the computing resources can not be guaranteed. This can be critical for an application of the contact model in an actual telepresence scenario where a stable operation with limit computing resources is required.

Another important aspect of the FEM is the discretization shall be done so that the validity of the governing system of equations in an element is preserved. If a property in an element is altered, the Lagrangian mesh must be reconsidered. This can be an issue for a complex mechanics problem which deals with multi discontinuities at the same time due to the mesh condition. A good example is the FEM analysis of hot cracking using temperature-dependent interface of SHIBAHARA ET AL. [77,78].

On the other hand, as an advantage, the remeshing provides a clear interface separation between continuous domains and a discontinuity. For an incision simulation, the edge of elements represent the edge of cut indicatively and require no post-processing for the visualization. The interaction with an external force asserted on a node of the edge of the cut can be treated as the Neumann boundary conditions of that current simulation step.

6.2 Extended finite element method

XFEM was originally developed to deal with a complex discontinuity with interface with a minimum consideration of the mesh conditions [14]. XFEM uses the extrinsic enrichments and the local version of partition of unity but apply the enrichment terms only on a certain set of local subdomains. The local subdomains are selected coinciding with the element containing cut.

The enrichment for the incision simulation is the Shifted function in this work. It endorses the internal dissection in an element with the local enrichment DoFs. As a result, the governing system of equations, is expanded to support additional DoFs as discussed in section 5.4. Although the expansion due to enrichment is done, the assembly process does not inquire a complete rebuild of the system matrices but only for the extended quadrants. The additional DoFs are allowed to be arranged after the highest index of the displacement vector instead of being as constituent variables among the original DoFs done by FEM with remeshing.

The remapping of boundary conditions is not necessary due to the unaltered Lagrangian mesh. The original nodes stay attached to the same material points since the initialization. It eases the treatment of boundary conditions significantly. These nature characteristics of XFEM approach provide an advantage in computing resource management. Because a discontinuity is systematically added to the system matrices, therefore the necessary computing resource can be calculated precisely. Also, if required, the already-added discontinuities can be taken out by neglecting the corresponding additional DoFs. Thus, the idea of enrichment enables the reversal of the system matrices to any previous configuration.

A disadvantage of XFEM is where the FEM with remeshing fulfills. The visualization of the edge of the cut does not explicitly demonstrated by XFEM and requires an extra post-processing. This is because the cut is treated in the local element domains and not in a global one. If the tetrahedron elements are small enough, textsc-Jeřábková uses the edge of tetrahedron elements subjected to the cut to visualization the approximated cut [46]. In this work, a linear is draw in the dissected elements as the visualization of the cut in the body as the scalpel is traveling through them. Moreover, if the interaction with the edge of cut is required, XFEM will meets its shortcoming if additional interpretation is not utilized due to lacking of true nodes on the edge of the cut.

6.3 Discussion on the practicality

The remarks in this chapter bring an essential discussion on the nature characteristics of FEM and XFEM approaches. In this section, the discussion will focuses specifically on the practicality of both approaches as a contact model for incision process in a telepresence application. Table 6.1 summarizes and evaluates both FEM and XFEM with respect to related categories.

Table 6.1: Evaluation of implementation practicality

	FEM	XFEM
Discontinuity implementation	o	+ +
Multiple discontinuities	- -	+ +
Dependency on mesh condition	-	o
Solving with Newmark implicit	-	+
Discontinuity visualization	+ +	-
Computing resource management	o	+

The implementation of discontinuity is where the XFEM gains advantages over FEM. With the local enrichment concept of the XFEM the discontinuity can be added only to the elements subjected to a discontinuity with minimal reconsideration of the mesh and topology. There are a lot of enrichment functions available for approximation of different types of discontinuities. XFEM allows one or more discontinuities to be existed in an element. In contrast, FEM depends strongly on the mesh. If the mesh condition does not suitable for a type of discontinuity, the simulation can not be achieved.

Using Newmark implicit integration requires a formulation of the dynamic governing system of equations. This is an disadvantage to FEM since the remeshing requires rearrange the DoF-indexes. The topology formulation of dynamic governing system of equations must be repeated at each simulation step. XFEM expands the governing system of equations systematically at its highest index. The formulation for integration does not require the initial part of the governing system of equations to be altered but only for the additional DoFs. This can result in a faster solving time.

The visualization of the boundary of discontinuities is a short-coming of XFEM. A post-processing is required to demonstrate the effect of the discontinuity such as the edge of the cut in the reference configuration. In contrast, the remeshing algorithm discrete the discontinuity domain clearly in reference configuration.

6.3 Discussion on the practicality

An ability to spontaneous model an unfamiliar object is a general requirement for a contact model in a telepresence application. The contact model must be able to simulate every actual contact mechanics in the remote environment. Moreover, the contact model should be efficient enough to guarantee its real-time simulation capability. This can be achieved with the necessary of hardware resource planed before hand. As discussed earlier, XFEM provides a range of flexibility how to model a contact mechanics with discontinuities. The maximum of memory required to store the dynamic governing of equations can be precisely calculated from the number of elements and the number of interested discontinuities. FEM depends on the remeshing algorithm to management the computing resources.

Considering XFEM is indeed developed from the conventional FEM, XFEM maintains the characteristic and the benefits of FEM but yet with extended capabilities. XFEM does prove the enrichment strategy suits better as a developing tool for a medical contact model because of its range of modeling flexibilities.

7 Adaptive empirical incision force model

A contact model is required to render both the visualization of the geometry deformation and the incision force for the haptic force feedback. FEM and XFEM discussed in the earlier sections demonstrate the geometry deformation of a solid body during incision indicatively. The incision forces can not be calculated from the governing system of equations derived from both approaches. Instead, they must be given as part of the Neumann boundary conditions acting on nodes of the dissected element where the scalpel is staying. Both FEM and XFEM require these boundary conditions to be solvable by the integrator. To render the incision force for the haptic feedback perception and to fulfill the required Neumann boundary conditions, an incision force model is empirically developed in this work based on the friction model in section 7.1.

However, the incision force involves in various variables and depends strongly on the situation, accurate rendered forces from a friction model may not be imminent while the friction coefficients remain constant throughout the incision process. In section 7.2, the adaptive parameter identification algorithm is developed. It verifies constantly the simulated feedback force to the actual force measurements from the force sensor and instantly adjusts the coefficients in the incision force model.

As the third algorithm, in section 7.3, an optimization algorithm is introduced to determine a set of optimum control parameters which guarantees a stable adaptation of the adaptive parameter identification algorithm.

7.1 Force modeling of haptic incision perception

The early implementation of force model for haptic force feedback rendering is the spring-damping such as the works of TERZOPOULOS ET AL. [91], JEŘÁBKOVÁ [46] and SELA ET AL. [74]. This model is simple and allow the human operator a sense of contact force. However, this implementation does not valid for the incision force due to an incorrect physical interpretation.

In Figure 7.1, the incision force $F_{Incision}$ is demonstrated to be the resisting force acting on the scalpel surfaces as the scalpel travels through the solid body in the incision direction. C denotes the current scalpel tip, which is pointed by vector r_C. Vector r_0 refers to the reference point where the incision begins.

In section 3, Figure 3.1, the definition of the cut in a solid-body is made, pointing that the normal vector n_t is the direction of the cut whereas the vector n_a is perpendicular to the cut direction and to the vector n_t. Therefore, the cut evolution

7.1 Force modeling of haptic incision perception

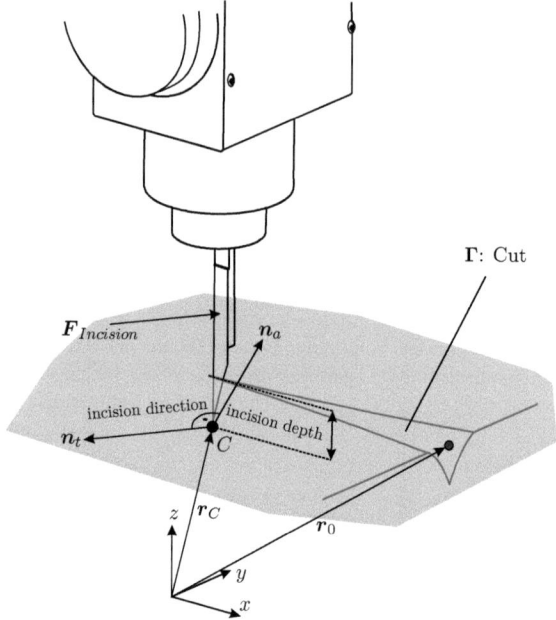

Figure 7.1: Incision force on the scalpel of the experimental teleoperator

is allowed only on the direction of the vector n_t and aligned with the cutting-edge of the scalpel on the end-effector of the telerobot. With these definitions at hand, there is valid following relations

$$F_{Incision} = \|F_{Incision}\| \cdot n_t \quad \text{and} \quad F_{Incision} \cdot n_a = 0 \qquad (7.1)$$

7.1.1 Friction on lubricated surface

In a survey conducted by ARMSTRONG-HÉLOUVRY [11], the friction force as a function of steady-state velocity of a pin on a flat metal plate is discussed. It is explained that friction is a physical process of shear in the junction changes with velocity. In the work of FULLER [30], various boundary lubrication conditions versus steady-state velocity are conducted. Figure 7.2 summarizes that limited and substantial boundary lubrication exhibits a large friction on the small velocity region because the lubricants provides only minimum or no boundary lubrication. In contrast, if the boundary lubrication between two metals is efficient, the friction in low velocity region will be small. All type boundary lubrications tend to provide a flat curve in the high velocity region where the partial fluid lubrication dominants. Therefore, the friction force is proportional to the velocity.

7.1 Force modeling of haptic incision perception

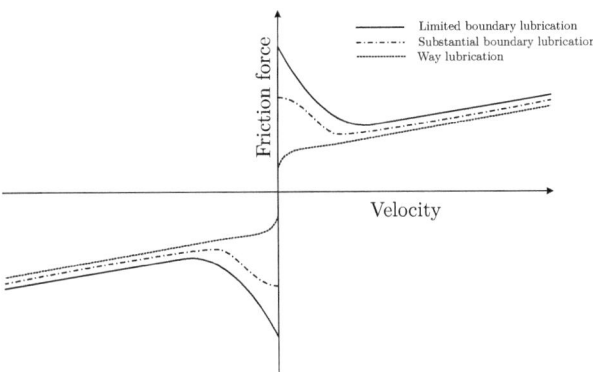

Figure 7.2: Friction as a function of steady-state velocity for various lubricants, adapted from Fuller [30]

Figure 7.2 is used as the basis physical description of the friction force occurs on the surface of the scalpel and the surface of the test object normal to the vector n_a. Because a medical incision is usually done on a tissue lubricated with fluid. The experiments done later in this work uses silicone as test objects. The nature of silicone provides a certain level of lubrication results in a slippery surface.

7.1.2 Incision force model in Cartesian system

The friction force from FULLER is a function of the steady-state velocity. To adopt this model, the friction force between surface of scalpel and the silicone should be assumed to proportional to the cut depth. In practice, a haptic device accepts the force signals in separate Cartesian coordinates, as well as the incision forces measured by the force sensor are given in a fixed Cartesian coordinate. Thus, it is coherent and convenient to refer the vector r_C and r_0 in a Cartesian system shown in Figure 7.1. The incision force model can then be described in a vector-component form as follow

$$F_{Incision} = - \begin{bmatrix} \tau_x \, \dot{x}_C \, (z_C - z_0) \\ 0 \\ \tau_z \, \dot{z}_C \, (z_C - z_0) \end{bmatrix} \qquad (7.2)$$

In this work, the force sensor and the haptic device share the same inertia Cartesian system and do not subject to rotational movement. Accordingly, the y-component of the incision force vector is always perpendicular to the incision direction n_t and aligned with normal vector n_a. This concludes that the y-component of the incision force model remains zero, since the incision does not exhibit in direction pointed by

7.1 Force modeling of haptic incision perception

vector n_a. The incision force model is a function of the cutting velocity ($\dot{x}$ or $\dot{z}$) depending on cut direction and the cut-depth ($z_C - z_0$).

With incision force model, the haptic force feedback rendering at a high-sampling rate is assuaged. The coefficients τ has unit of Ns/m^2 have the most crucial role for a precise incision force prediction. It is related not only to the friction coefficient but also depends on the geometry of the employed scalpel. For example, τ_x relates to the cutting edge of the scalpel, whereas τ_z to its the sharp end. If a scalpel with a sharp end similar to one in Figure 7.1 is employed, it is assumable that τ_z should smaller than τ_x since the object introduces a smaller resisting force as it is being pierced.

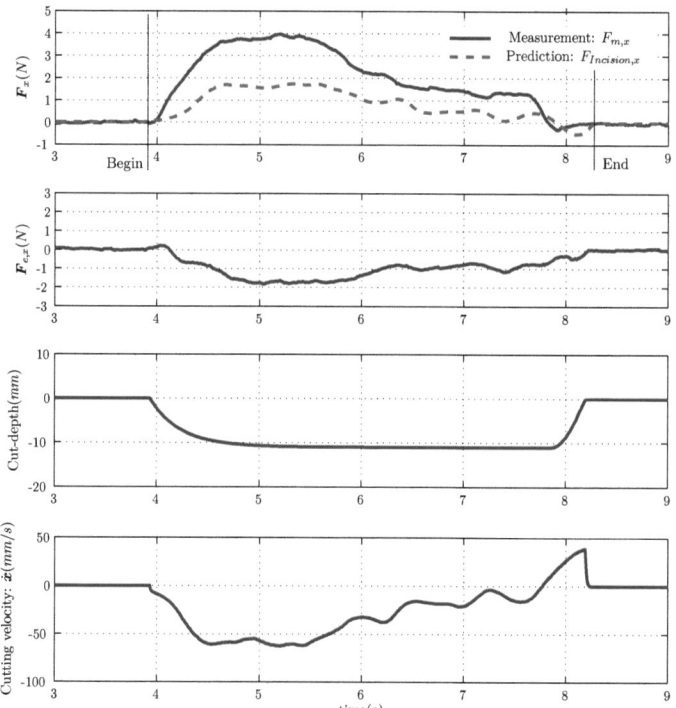

Figure 7.3: Comparison of the measured and the predicted incision forces in horizontal direction align with the x-direction with constant $\tau_x = -2 \times 10^{-3} \frac{Ns}{m^2}$

For an explanation of the empirical incision force model, Figure 7.3 and 7.4 shall be discussed. Both demonstrate the results obtained from an experiment which undergoes an incision process in a silicone test object. The measured incision forces in x- and z-directions were obtained using a force sensor at the end-effector of the teleoperator. Simultaneously, the empirical incision force model from Equation 7.2

7.1 Force modeling of haptic incision perception

computed the incision force according to the scalpel position in the silicone test object with the arbitrarily constant coefficients $\tau_x = \tau_z = -2 \times 10^{-3}\, Ns/m^2$. The incision began when the tip of the scalpel pierced through the surface of the test object which can be noticed from the change in cut-depth. The scalpel was therefore drawn in the negative x-direction to exhibit a cut in the test object with an almost constant cut-depth before ended by pulling the scalpel out of the test object.

In Figure 7.3, the measured incision force $F_{m,x}$ is plotted against the cut-depth and cutting velocity. It is obvious, as proposed, that the measured incision force is proportional to the cut-depth and the cut-velocity as the scalpel travels through the object in $-x$-direction. On the other hand, the predicted incision force $F_{Incision,x}$ provides a similar result with the aspects of incision process are presented correctly. However, a smaller magnitude is noticed due to an incorrect value of coefficient τ_x. The force error $F_{e,x}$ reveals that the error between the measured and the predicted forces is not constant during the incision process which points that coefficient τ_x can not be a constant. This can be explained using the friction-velocity curve from Figure 7.2 that the friction coefficient is not proportional throughout the velocity region. The horizontal incision force in x-direction is foremost important for the haptic perception, since it demonstrates the haptic perception of the resisting force as the scalpel travels through the test object.

In Figure 7.4, the empirical incision force model was again employed to predict the incision force for the z-direction of the same experiment as previous. The z-component of the incision force is mostly noticeable when the cut-depth is changed or during piercing and pulling the scalpel out of the test object. This vertical incision force perception is especially critical during the first impact with the test object. Similar as of previous, the incision force model demonstrated the piercing force occurring during the first impact of the sharp end of the scalpel with the test object as well as the friction force acting on the scalpel surface as the scalpel was pulled out. The magnitude of the predicted incision force in z-direction $F_{Incision,z}$ was larger than the measured one. This convinces that the given coefficient τ_z is obviously inaccurate. In the same manners as of the x-direction, the force error $F_{e,z}$ was not constant pointing to the fact that the coefficient τ_z shall not remain as a constant throughout the incision process.

A teleoperator is often controlled by hands of its human operator using a haptic device. An inaccurate coefficient τ can cause the human operator a false decision and damages the remote environment. An accurate perception without a distortion is therefore required. However, actual telepresence applications are usually appointed to work in a unknown environment and conditions. Therefore, for a medical telepresence system, the incision is mostly done to an object which its characteristic may be unknown. Accordingly, the coefficient τ in the incision force model must be assumed to be unknown in the first place. τ can be theoretically predetermined for each material and object found in the remote environment. Nevertheless, this off-line parameter identification may not be practical or functional in the environment which changes rapidly or there is an uncertainty in the material type. An adaptive parameter identification algorithm, discussed in the next section is one of a logical

7.2 Adaptive parameter identification

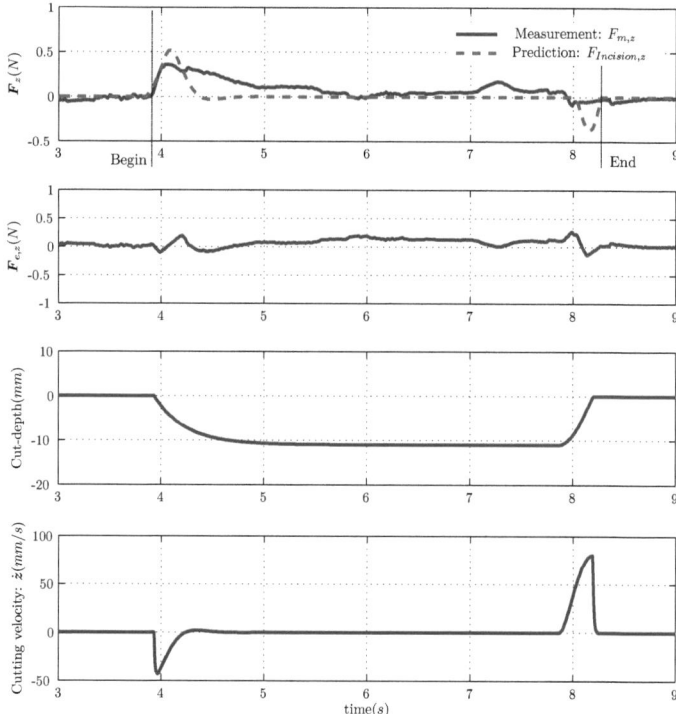

Figure 7.4: Comparison of the measured and the predicted incision forces in vertical direction, z direction with constant $\tau_z = -2 \times 10^{-3} \, \frac{Ns}{m^2}$

solution providing a continuous verification capability to the incision force model and reconfiguration of the coefficient τ in each direction.

7.2 Adaptive parameter identification

A telepresence system with a contact model is discussed in term of control theory as a bilateral force-position control architecture with time delay in the communication medium. The Smith predictor is a contact model in telepresence applications. The Smith predictor may achieve a successful compensation of time delay if the contact mechanics are modeled without a flaw. At this ideal state, the actual feedback signals from the teleoperator in the remote environment are negligible while the stability of the telepresence system is maintained.

In the last section, it was discussed that the model of the contact mechanics between

7.2 Adaptive parameter identification

teleoperator and its remote environment must be determined before the initialization of a teleoperation. However, the correct modeling and identification of the environment are achievable only in some particular situations and not in general. In fact, it is arguable that the concept of Smith predictor discussed in chapter 2 can actually be realized in a genuine telepresence application and scenario. As a defensive argument, in this work, the incision force model in equation 7.2 can be continuously validated during the actual incision.

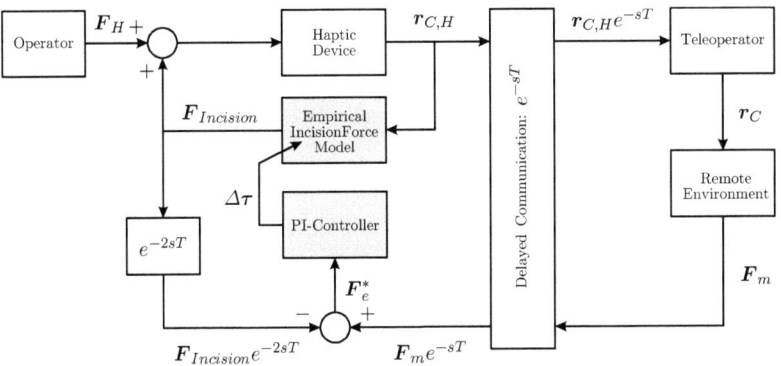

Figure 7.5: Adaptive incision force model algorithm in a telepresence system

Figure 7.5 demonstrates a proposed control loop for a telepresence system. The teleoperator is deployed to the remote environment whereas its human operator controls it from a local station using a haptic device. The human operator actuates the haptic device providing a force vector $\boldsymbol{F}_H$ allowing the haptic device to register the reference position of the teleoperator end-effector $\boldsymbol{r}_{C,H}$. This position commands are passed through a delayed communication with an assumed constant time delay T to the teleoperator. The movement of the teleoperator is depicted by the actual position of its end-effector $\boldsymbol{r}_C$. The contact and interaction with the remote environment provides a contact force $\boldsymbol{F}_m$ which is measurable with a force sensor. $\boldsymbol{F}_m$ is required to be fed back via the same delayed communication to the local station where the human operator exists. For a conventional telepresence system, due to time delay of the close-loop, the arrival of $\boldsymbol{F}_m$ at the human operator will be $2T$ later after the reference position $\boldsymbol{r}_C$ is sent. Therefore, the time delay leads to an unsynchronized force-position relation at the human operator. The phenomena results in a unrealistic experience which may cause the human operator a false decision because of the reaction to the feedback with phase shift.

There are some implementations of Smith predictor which do not neglect the delayed feedback signal. On the other hand, the proposed control loop does not allow the

7.2 Adaptive parameter identification

controller to respond to the delayed feedback. The reason focuses on the large time delay. The phase shift between $r_{C,H}$ and F_m can be emphasized to be more than $180°$ high frequency movement. Thus, feeding F_m to the operator can significantly reduce the teleoperation performance.

For this reason, F_m is necessary for the contact model validation purpose for the proposed control architecture. Doing so, an adaptive parameter identification is integrated as a subroutine to the empirical incision force model. Two essential design criteria of the adaptive parameter identification were taken into account:

- The algorithm must instantaneously respond to the force errors F_e^* and effectively provide accurate coefficients τ for every incision direction during the actual incision is being carried out.

- The convergence of the coefficients τ must be guaranteed yet provide a transparent transition on the force perception of the human operator.

The incision force model is initially given a set of coefficients τ_0 for each incision direction. These coefficients may be collected from a prior experiment or an available contact library. The adaptive parameter identification keeps validating the predicted incision force $F_{Incision}$ to the actual measurements F_m. To validate the predicted force signals $F_{Incision}$, an artificial time delay $2T$ must be included purposefully allowing the phase synchronization with F_m. The force errors F_e^* between the two force signals are then obtained and provided to the adaptive parameter identification algorithm for the determination of correcting coefficients $\Delta\tau$. Assumed that the teleoperator follows the position commands from the human operator perfectly, $r_{C,H} \cong r_C$, and the linear contact model is adequate to accurately approximate the environment mechanics, this architecture can achieve an improved performance even when the time delay in communication is varied. However, the actual amount of time delay in the communication must be known. A similar approach is proposed by FITE ET AL. in [28] which adapted from the work of SLOTINE ET AL. in [82]. The other interesting adaptive parameter identification algorithms are also discussed by SMITH ET AL. in [84]. The approximated time-dependent coefficients $\tau(t)$ can be computed as

$$\tau(t) = \tau_0 + \int_{t_0}^{t} \mathrm{d}\tau(t) \tag{7.3}$$

or in discrete form

$$\tau(t_{n+1}) = \tau_0 + \sum_{n=1} \Delta\tau(t_n) \tag{7.4}$$

To satisfy these design criteria, a PI-controller depicted in Figure 7.6 is chosen as an adaptive parameter identification. A set of control gains K_P and K_I can be chosen to provide an optimum and convergent result required by the design criteria.

To demonstrate the parameter validation in the incision force model, the incision process identical to that of Figure 7.3 and 7.4 is brought into discussion in this section.

7.2 Adaptive parameter identification

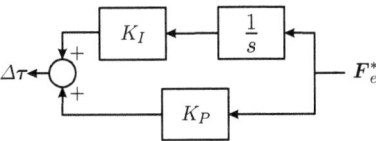

Figure 7.6: PI-controller responds to the force error F_e^* and provides the correcting coefficients $\Delta\tau$

An artificial time delay of 150 ms was given representing the delay in communication channel. For the horizontal x-direction, the coefficient $\tau_{x,0} = -2 \times 10^{-3}\ Ns/m^2$ was given initially as previous.

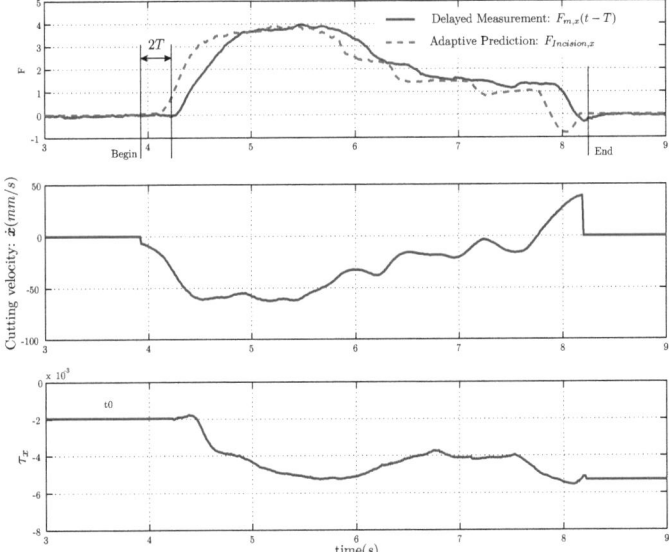

Figure 7.7: Non-delayed predicted incision force $F_{Incision,x}$ with adaptive coefficient τ. Artificial time delay $T = 150\ ms$

In Figure 7.7, the predicted incision force $F_{Incision,x}$ was calculated instantaneously based on the value of $\tau_{x,0}$ regardless of time delay. $2T$ later, the measured incision force $F_{m,x}$ arrived, the adaptive parameter identification began to validate the accuracy of the incision force model from the force error $F_{x,e}^*$. The value of coefficient τ_x was accommodated from the surmised initial value. The correct value of the coefficient τ_x was achieved when the force error was zero providing an accurate predicted incision force. Nevertheless, since the incision force is subject to the cutting velocity

7.2 Adaptive parameter identification

$\dot{x}$, the coefficient τ_x remained being adjusted until the end of the scalpel penetration in the test object. The obvious difference of coefficient τ_x in high and low cutting velocity regions is demonstrated which confirms Figure 7.2.

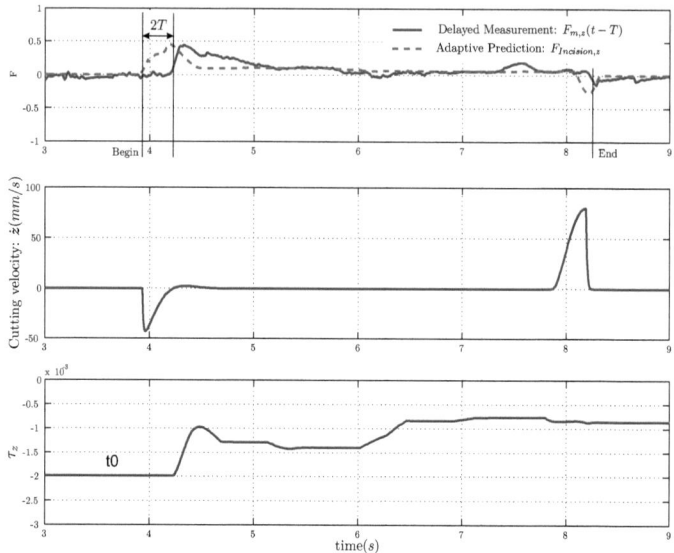

Figure 7.8: Non-delayed predicted incision force $F_{Incision,z}$ with adaptive coefficient τ. Artificial time delay $T = 150\,ms$

Similarly, in Figure 7.8, the incision force in vertical z-direction was predicted with an identical initial value of coefficient $\tau_{z,0} = -2 \times 10^{-3}\,Ns/m^2$. The incision force model reacted on the collision of the scalpel tip with the test object and calculate an amount of the feedback force using predefined $\tau_{z,0}$. The adaptive parameter identification began to correct to obtain the actual value of the coefficient τ_z as stated by equation 7.3 and 7.4. The final value of τ_z is not identical and smaller than that of τ_x which is consistent with the assumption that τ is scalpel-geometry dependent. Hence, τ for each incision direction can be varied.

For a discussion on the accuracy improvement of the incision force model, the predicted incision force $F_{Incision,x}$ is purposively delayed by $2T$ in Figure 7.9. This allows a direct magnitude comparison of the predicted incision force and the measurement. $F_{Incision,x}$ is apparently more accurate with adaptive parameter identification algorithm when compared to the earlier results without adaptation in Figure 7.3.

7.2 Adaptive parameter identification

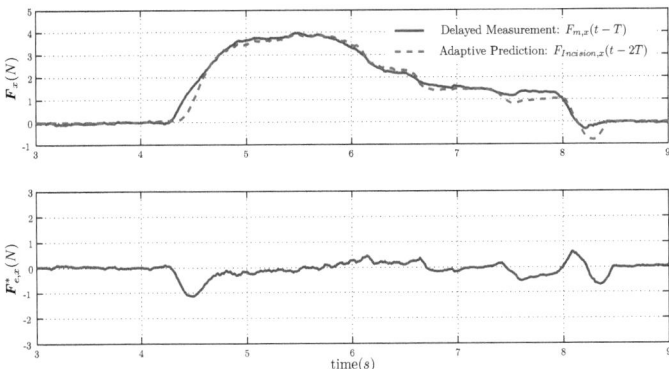

Figure 7.9: Incision force error $F^*_{e,x}$ shows an accuracy improvement with adaptive parameter identification algorithm in the horizontal direction

In the same manner, Figure 7.10 demonstrates an improvement of the prediction over one without the adaptive parameter identification algorithm depicted in Figure 7.4. This is a useful benefit of having the adaptive parameter identification algorithm since it allows an approximate model parameter to be existed in the beginning while the correction can be made when a necessity arises. The use of PI-controller as the adaptive parameter identification provides a transparent correction to the force feedback rendering while the natural presentation of the incision force is preserved.

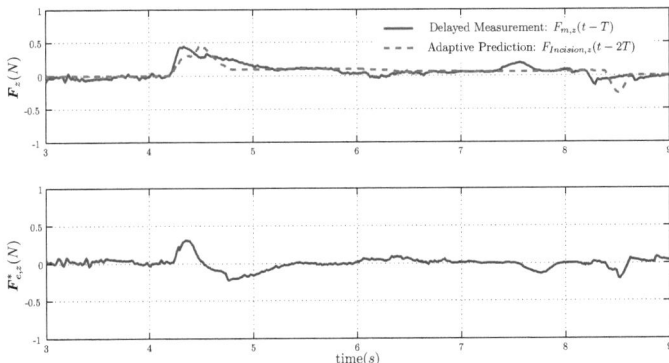

Figure 7.10: Incision force error $F^*_{e,z}$ in the vertical direction also declines when the adaptive parameter identification algorithm is employed

7.3 Optimization

Safety and stability of the teleoperator in the remote environment gain the highest priority in designing the controlling strategy. Since the alternation to the coefficients τ by the adaptive parameter identification affects the force perception of the human operator directly, the algorithm does not allow to exhibit an oscillation in predicted incision forces. The damages to the remote environment is therefore inevitable as a consequence of a potentially false decision of the human operator.

The performance of the adaptive parameter identification differs from the given set of gains in PI-controller $p = \{K_P, K_I\}$. It is the fact that the PI-controller can be more responsive with a higher gain-setting. However, with inappropriate gains, an strong oscillation as depicted in Figure 7.11 can be provoked.

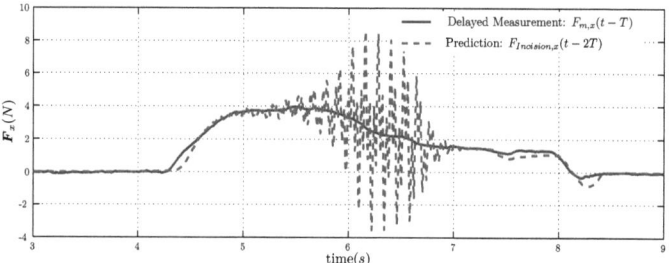

Figure 7.11: Inappropriate values of gain K_P and K_I lead in a strong oscillation in force feedback

For this reason, a set of optimized controller gains p for each individual incision direction is compulsory and must be determined prior to initiation of the adaptive parameter identification algorithm and an actual incision process to guarantee a stable and natural iteration of coefficient τ. In this work, the performance of the adaptive parameter identification is assessed by the total amount of force error F_e exhibited when a set of controller gains during an incision is used. An optimum p is found when the total amount of force error is minimum. For this purpose, the following optimization criteria is employed.

$$\min_{p} \int (F_e^*(p))^2 \, dt \quad \text{with} \quad p = \{K_P, K_I\} \tag{7.5}$$

This optimization criteria requires a complete set of measurement of incision forces and movement of the haptic device to be efficient. This is reasonable since the incision process with teleoperator is complex and non-linear with human-factor included. If the incision process and the telepresence system is assumed to be a linear system, an uncertainty in the position of the system poles will occur.

The Nyquist stability analysis may not be an efficient option to determine the input frequency region causing instability of the incision contact mechanics because of model parameter uncertainty [38]. Hence, if only a fragment of the measurement is considered, a classical controller gain selection based on stable pole placement may lead to an certain set of p which is not suitable for a specific range of frequency of input similar as depicted in figure 7.11.

However, the determination of the controller gains can be done off-line using the measurement collecting from an incision process with the telepresence system. Moreover, it is found from the experiments that the optimization algorithm is not frequently required for similar material of test objects. The set of optimum controller gains provide a robust performance for handling the incision process with test objects made of silicone of different coefficient τ.

7.4 Implementation of the adaptive contact model

With the adaptive incision force model integrated into the geometry deformation simulation algorithm introduced earlier chapter, the adaptive contact model for the incision process is completed. The Neumann boundary conditions for FEM and XFEM dynamic geometry deformation simulation will be fulfilled by the calculated incision force.

The initialization of the adaptive contact model is depicted by algorithm 5. The dynamic geometry deformation simulation must be initiated depending on the chosen simulation approach. Algorithm 1 is for FEM with remeshing and Algorithm 3 for XFEM. The initialization of the empirical incision force model requires a set of coefficients τ_0 specific for the material of the test object and the geometry of scalpel. To determine the cut-depth, the reference level z_0 of the surface of the simulated soft body must be given. For the adaptive parameter identification algorithm, a set of optimized control gains K_p and K_I is required for instantaneous correction of the modeling error in the incision force model.

Algorithm 5: Initialization of the adaptive contact model

Initialization geometry deformation simulation: FEM or $XFEM$
Determine incision force model:
  $\tau_0, z_0 \longrightarrow$ equation 7.2
Adaptive parameter identification:
  $optimized\ K_P, K_I$
Simulation step indicator:
  $t_n = 0 \longleftarrow n = 0$

The adaptive contact model is implemented as depicted in algorithm 6. The scalpel position is obtained as usual from the employed haptic device. The geometry of

7.4 Implementation of the adaptive contact model

the considered solid body for the next simulation step U_{n+1} is calculated as demonstrated in Algorithm 2 for FEM approach or Algorithm 4 for XFEM approach. The incision force for the next simulation step $F_{Incision,n+1}$ is obtained from the empirical incision force model using the current coefficients τ_n. $F_{Incision,n+1}$ justifies the Neumann boundary conditions for the dynamic geometry deformation on the dissected element.

The adaptive algorithm validates the predicted incision force to the actual delayed incision force from the force sensor, if an error is exhibited, the adaptive parameter identification algorithm reacts by providing $\Delta\tau_n$ which is used to update the empirical incision force model for a more-accurate incision force prediction at the next simulation step.

Algorithm 6: Adaptive contact model for incision process

---

for $Simulation = true$ **do**
  Read current scalpel position: $\longrightarrow r_{C,n}$
  Geometry deformation simulation: $\longrightarrow U_{n+1}$
  if $r_{C,n}$ contacts with soft body **and** $\dot{r}_{C,n} \neq 0$ **then**
    Predict incision force: $\longleftarrow \tau_n$
    Update incision force: $\longrightarrow F_{Incision,n+1}$
    Read delayed actual incision force: $\longrightarrow F_{m,n-T/\Delta t}$
    if $F_{Incision,n-2T/\Delta t} \neq F_{m,n-T/\Delta t}$ **then**
      Adaptive parameter identification: $\longrightarrow \Delta\tau_n$
      Update incision force model: $\tau_{n+1} = \tau_n + \Delta\tau_n$
    end
  end
  else
    Update incision force: $\longrightarrow F_{Incision,n+1} = 0$
  end
  Update simulation time: $n = n+1$, $t_{n+1} = t_n + \Delta t$
end

The benefit of the proposed adaptive contact model is its ability to achieve a realistic simulation of a soft body when deforms during an incision process. The adaptive contact model can be employed as a stand alone algorithm enabling a virtual experience as a surgical simulator.

Although the mesh generator in this work is developed specifically to provide a mesh valid for the silicone test object. There are many open-sourced mesh generators, such as TETGEN [79, 80] which except the industry standard object format such as *obj* and *3ds*. A complex object such as a three-dimensional geometry of a real organ obtained from *Computer Tomography* or Magnetic Resonance Imaging often adopt

7.4 Implementation of the adaptive contact model

this formats. Therefore, the contact model in this work can be employed to simulate the deformation of these objects subjecting to an medical incision.

The reference level of the surface z_0 is treated as a constant in this work. This shortcoming can be solved with an array of sensor. In the actual teleoperation, a camera system is often available. The use of image processing technique can provide the level of the surface of where the incision is about to take place. Another possibility is to use a distance sensor to measure the actual surface level. These practical can enhance the practicality of the adaptive contact model in this work.

Moreover, the adaptive contact model is expandable to include the other discontinuities of interest without a complete revision. This is possible since XFEM approach is implemented in the contact model. The discontinuity approximation terms can be added locally to each affected elements.

8 Experiments and results

This chapter begins with the discussion on the experiments employing the adaptive contact model developed from chapter 3 to 7. At first, the experimental telepresence system and the remote environment will be discussed in section 8.1. Second, in section 8.2, the experiments on an instability from a large time delay during incision processes will be investigated. In this section, the adaptive contact model developed in this work will be demonstrate to compensate the time delay while improve the stability of the telepresence system during incision.

8.1 Experimental telepresence system

The telepresence system employed in this work is depicted in Figure 8.1. The system comprises with two essential elements, which are the teleoperator in the remote environment and the local station used by the human operator to manipulate the teleoperator.

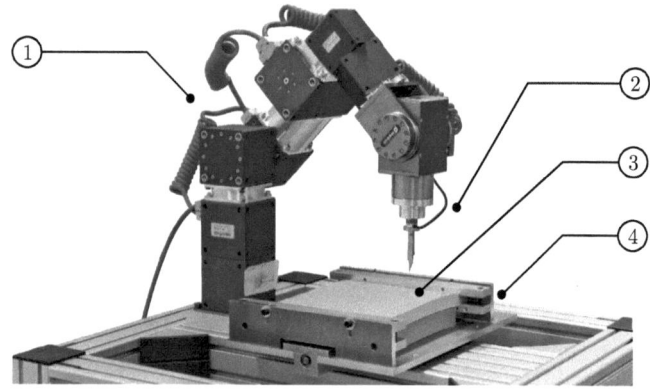

1. 6-DoF modular robot
2. Force-torque sensor with scalpel
3. Test object
4. Tension platform

Figure 8.1: Experimental telepresence system

The teleoperator employed in this work is set up in a standard 6-DoF industrial robot configuration. This configuration is chosen for its adequate work space and

8.1 Experimental telepresence system

its flexibility to perform an incision in the test object from above. Between the teleoperator effector and the scalpel, a 6-DoF force-torque sensor is attached. It measures the actual incision force which is either transfered to the operator via haptic device with or without time delay. In addition, as the remote scenario imitating the abdominal surgery, a tension platform is developed to mount a silicone test object in place while providing a tension force to the test object causing its geometry deformation.

1. Visualization of the remote enrivonment   2. Haptic device

Figure 8.2: Local station

An operator on the local station is allowed to control the position of the scalpel via a dedicated haptic device. A PHANToM Omni haptic device from SENSABLE [7] is connected to a PC providing the position command of the end-effector to the teleoperator. As the operator interacts with the telepresence system, the bilateral control loop of system is closed while the operator responds to the force feedback rendered by the haptic device.

The adaptive contact model developed in this work runs on the PC unit at the local station. Since in the remote environment, there is no camera system, the operator manipulates the teleoperator relying on the visualization of the remote environment provided by the adaptive contact model. The force feedback can be obtained from the force-torque sensor or simulated using the adaptive contact model.

A selectable artificial time delay is implemented to postpone the position command from haptic to the teleoperator and the force signal from force-torque sensor to the haptic device. It is necessary due to the actual distance between the remote environment and the local station does not large enough to provoke a desired large time delay.

8.1.1 Teleoperator

The teleoperator comprises with 6 SCHUNK/AMTECH [6] modular actuators employing CAN-standard as a communication protocol. The modulars are serial connected with dedicated cable providing necessary communication and power. An emergency switch is hard-wired to the power supply unit. It is capable of forcing the actuator modules to automatically engage the mechanical break holding the teleoperator in place. The teleoperator is connected to the CAN-master housed in a PCI-e slot of the PC. A data-acquisition card is required for reading the force signal from force-torque sensor.

In order to control a robotic system or a manipulator, its forward and inverse kinematics must be thoroughly investigated. The employed robotic system is known as a serial kinematic with its base fixed relative to an inertial frame. Its end-effector at the another end is allowed for manipulation of an object in workspace. Figure 8.3 depicts the teleoperator stays in the inertial frame 0. Its end-effector processes C coordinate. The position of the end-effector is pointed by a vector r_C. The orientation of the scalpel coincides with the orientation of the end-effector. It is measured by the relative angles of C coordinate with respect to inertia frame 0.

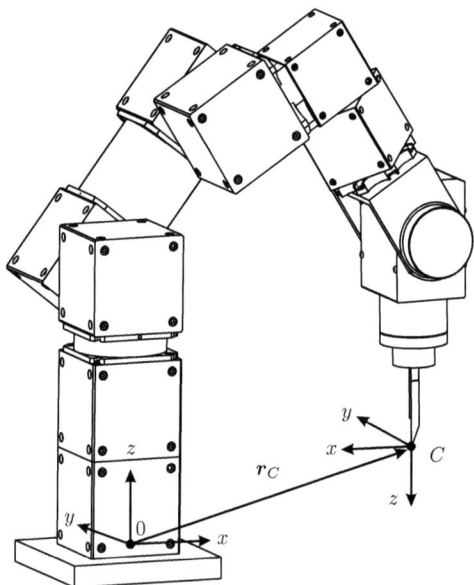

Figure 8.3: Kinematic scheme of experimental teleoperator

The vector $\bm{q} = [\; q_1 \; q_2 \; q_3 \; q_4 \; q_5 \; q_6 \;]^T$ contains all available minimal coordinates of the teleoperator. The position and *Kardan*'s angle of the manipulator end-effector can be written together in a vector form as $\boldsymbol{w} = \begin{bmatrix} {}_0\boldsymbol{r}_C^T & \alpha & \beta & \gamma \end{bmatrix}^T$ or as the forward kinematic with respect to its minimal coordinates $\boldsymbol{q}$ as

$$\boldsymbol{w} = f(\boldsymbol{q}) \tag{8.1}$$

In contrast, the corresponding minimal coordinate $\boldsymbol{q}$ must be found for a given vector $\boldsymbol{w}$ in order to control the individual actuators. Thus, the inverse kinematic in Equation 8.2 must be constantly calculated.

$$\boldsymbol{q} = f^{-1}(\boldsymbol{w}) \tag{8.2}$$

It must be aware that a unique solution to inverse kinematic may not always exist. For a further extensive discussion on solving the inverse kinematic of the employed teleoperator, the interested reader is encouraged to consult [27]. Some interesting relating examples can also be found in [17, 18, 40, 67, 81].

8.1.2 Tension platform

As a remote environment, a tension platform shown in Figure 8.4 is developed. The dimension of the silicone test object lying on this platform resembles area common for an abdominal surgery.

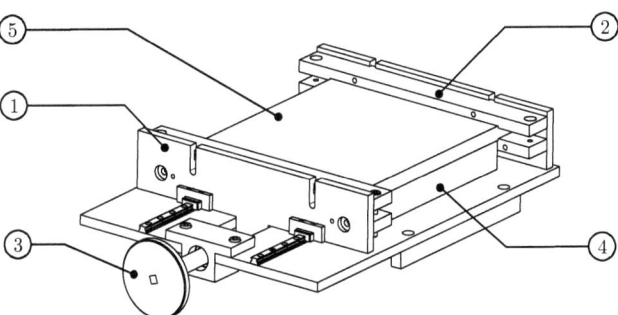

1. Length adjusting gripper
2. Second gripper
3. Length adjusting knob
4. Silicone support: SI60GB
5. Test object: SI6GB

Figure 8.4: Tension platform (remote environment)

The essential design criteria of the tension platform is the uniform deformation distribution on both sides of the test object. The platform has two grippers hold both sides of the test object in place. The length-adjusting gripper sits on two rail guides. A knob on a side of the platform is connected to an axial screw spindle allowing a length adjustment between grippers. This mechanism allows the test object to split evenly when subjected to a cut similarly to abdominal surgery. Whereas the boundary condition around the grippers can be given as in term of deformation instead of applied forces. This ease the treating of boundary condition considerably.

Therefore, the length adjustment coincides with the Dirichlet's boundary condition and must be given during initialization of the simulation as part of the initial boundary values. The mechanical design of the platform provides a minimum exhibition of the mechanical looseness. Hence, the length between grippers remains unchanged and allow the deformation simulation to treat a constant Dirichlet's boundary condition throughout the simulation.

Additionally, the test object made of soft silicone is placed on a hard silicone support made of silicone SI60GB (Appendix A,Table A.1). This support is employed to compensate the push force provided by the teleoperator as the incision is being initiated. It allows the under part of the test object to remain still while avoiding the collision between scalpel and a solid metal part of the platform. Since its shrinkage stays well <0.1%, the height of this silicone support can also be treated as constant Dirichlet's boundary conditions for the simulation.

8.1.3 Test object

Biomechanic properties of human skin have been studied by AGACHE ET AL. [9]. Although the work concluded that the elasticity of human skin is age dependent, it did provide an important evidence showing the Young's modulus of average human's skin stay in the range of $4.2 \times 10^5 \ N/m^2$ to $8.5 \times 10^5 \ N/m^2$. Silicone SI6GB is the material of test objects employed for every experiment done in this work. It is chosen for its similar elasticity to the human skin and widely used as a material for artificial organs for medicine study and demonstration purposes.

In order to validate the SI6GB as the test object, its Young's modulus must be determined and verified. As depicted in Figure 8.5, the tension platform was modified allowing the axial tension force measurement with a force sensor attached to one of the gripper. The total axial deformation was provided by a laser profile scanner. The laser profile scanner was mounted on the teleoperator end-effector for precise positioning.

The Young's modulus E was then determined with an assumption that the Poisson ratio ν is 0.49 for a material with elastic behavior similar to rubber [62]. The measurement of tension forces and total deformation from the experiment with the test object were compared with the result given by the static FEM simulation. The actual value of Young's modulus was found by iteration the value of Young's modulus until the simulation results identical to the measurement. The value of Young's modulus was found as $3.85 \times 10^5 \ N/m^2$ which has a coincided elasticity similar to human skin. The density value of the silicone SI6GB ρ was calculated by the total mass per liquid volume as stated in Appendix A Table A.2. The value of Young's

8.1 Experimental telepresence system

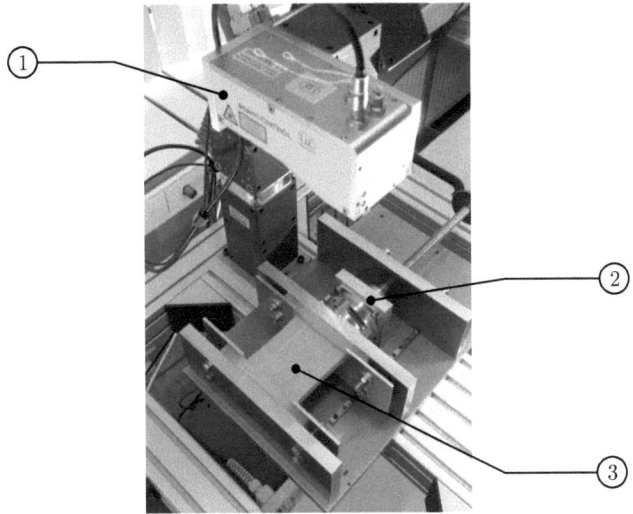

1. Laser profile scanner   3. Test object: SI6GB
2. Axial force sensor

Figure 8.5: Verification of Young's modulus of experimental silicone

modulus E, Poisson ration ν and material density ρ can be used for simulation with FEM and XFEM as material parameters.

8.1.4 Software framework

The software development in this work is multi-thread coding. In computer science, the term threaded code refers to a compiler implementation technique where the generated code essentially consists entirely of calls to subroutines. The code may be processed by an interpreter, or may simply be a sequence of machine code call instructions. One of the main advantages of threaded code is that it is very compact, compared to code generated by alternative code generation techniques and alternative calling conventions. This enables programmers to concentrate mostly on the structure of the software framework and synchronization mechanism between subroutines and program modules.

Windows XP 32-bit is employed as the operating system and developing environment of the software package in this works. This operating system is not a real-time operating system which the exact execution time of a command can not be given. However, it allows multi-threading and ability to set the priority of each executable program modules. Thus, the software framework was developed by considering a proper hierarchy of the program modules priority. The software framework comprises with 3 important program modules

1. Geometry deformation simulation and visualization
2. Haptic interface and Adaptive empirical incision force algorithm
3. Central teleoperator kinematic control

The haptic interface gains the highest priority due to its demanding a high-sampling rate of force feedback generated by the adaptive empirical incision force algorithm. The central teleoperator kinematic control has the second highest priority, since a smooth movement of the telerobot can be achieved by updating the position command at lower sampling of $100\,Hz$. This is possible because the actuator modules have cascade controller built-in and responsible of the path generation at the minimal coordinate level. The geometry deformation simulation with the sampling rate of visualization as low as $30\,Hz$ is sufficient while the animation remains seamless. Therefore, it is given the lowest priority. An open-source software package CHAI3D which is well-known for haptic applications [24] is employed to associate the development of this software framework. This software package provides a unified haptic interface to some of widely accepted haptic devices available. Its visualization is implemented based on standard OPENGL API [4]. Although the haptic feedback rendering is also extensively available by the package, it does not have incision force rendering capability. Therefore, the incision force calculation is done solely relying on the implementation proposed in this work.

8.2 Time delay compensation utilizing contact model

The analysis of the effect of time delay is essential for the utmost understanding of the instabilities during a telepresence operation. The typical linear evaluation methods based on root locus and bode diagram are not a proper investigation tool to determine the instability during an actual medical incision due to the complexity of modeling and non-linearity of the incision contact mechanics. For a comparison purpose, a large time delay causes reduction in gain-margin and phase-margin from a linear control theory. Instead, the investigation and discussions of the impact of time delay in this section was done based on three experiments.

Section 8.2.1 comprises with two experiments which are designed to address the issues of time delay during incision process with a telepresence system. In Section 8.2.2, with the third experiment, the utilization of adaptive contact model developed in this work will be discussed as a solution to the time delay problem.

8.2.1 Impact of time delay during incision

Reference experiment without time delay

The first experiment is the reference aiming to an operator who have less experience with manual robot control and telepresence system. It also provides the operator a picture of an ideal and transparent telepresence experience. An operator is allowed to perform one incision on a single xz-plane to the silicone test object (see Figure 8.3 for coordinate notation). The incision force was provided by the force-torque sensor at the end-effector of the teleoperator and forwarded directly to the employed haptic device. The force signal was not delayed by the artificial time delay.

Figure 8.6 shows the incision force in horizontal direction x together with the cut-depth and the cutting velocity. The first impact as the scalpel pierced through the test object was recognized by the synchronized incision force signal. As the incision process was being carried out, the hand's movement was synchronized with the force feedback until the end of the process. The magnitude of the incision force was also exhibited as the function of cutting velocity as discussed in section 7.1.2.

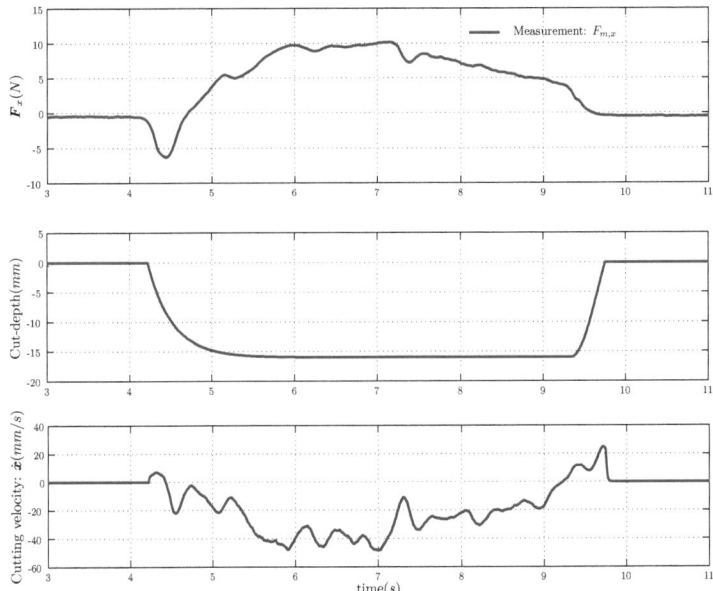

Figure 8.6: Stable telepresence operation with synchronized hand's movement and incision force

This ideal telepresence experience was reflected by Figure 8.7. The quality of the cut with no damage exhibited in the test object depicted is achieved. It also corresponded with the report of minimum psychological interference during the incision where the operators could perform the incision naturally relying on their perceptional individual. The further trials of this experiment also showed an improvement of the operator's confidence on using telepresence system.

Figure 8.7: The cut obtained from a stable telepresence operation exhibited no damage

With 300 ms time delay

The second experiment is designed to provoke the instability using a predetermined artificial time delay. This time delay was set as $150\,ms$ or $300\,ms$ for bilateral communication. The latter of experiment variables remain identical to the earlier reference experiment without time delay. Figure 8.8 illustrates the instability in the telepresence system due to postponed arrival of the incision force signal.

The operator encountered a false force perception during incision. This phenomenon deteriorated the ability of task handling, thus reduced the confidence of the operator. The missing of necessary force feedback perception caused an uncertainty to the operator regarding the cut depth of the first contact with the test object. When the force signal of the first contact arrived, it did not represent the quantity of piercing force. The unsteady movement of the hand especially in handling the cut-depth were caused by series of delayed and unsynchronized force feedback. A difficulty to analyze the situation while perceived an incorrect force feedback was obvious in this experiment.

The cut result shown in Figure 8.9 is an example of damage caused by adverted instability to the test object. The strong oscillation in cut-depth was the cause of the jagged cut edge.

8.2 Time delay compensation utilizing contact model

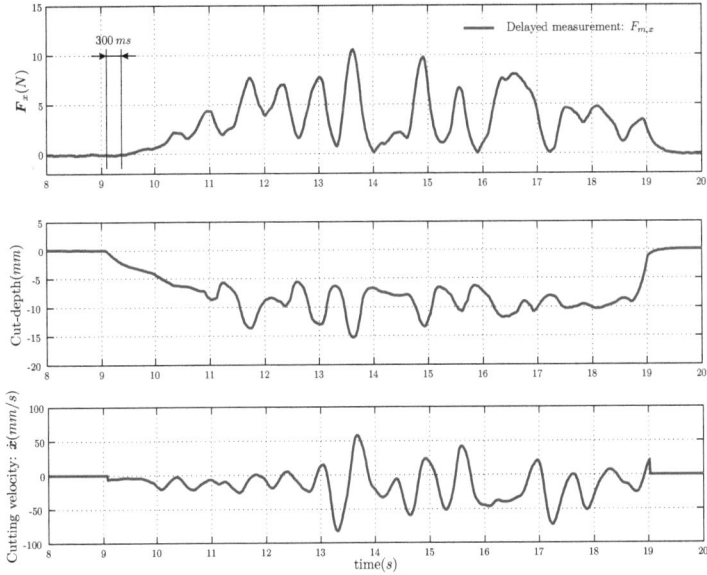

Figure 8.8: Unstable incision with a strong oscillation in cut-depth due to non-synchronization between hand's movement and force feedback perception

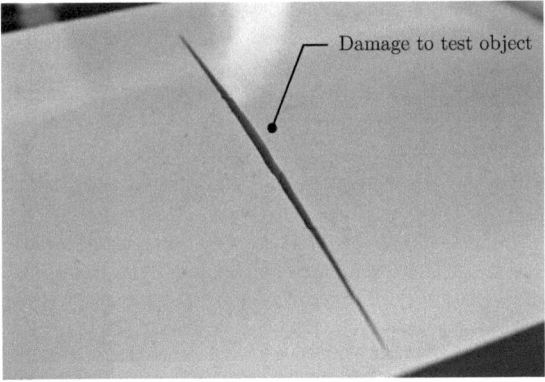

Figure 8.9: With $300\,ms$ time delay, the instability caused damage to test object presenting a jagged cut edge.

8.2.2 Incision force compensation utilizing contact model

This section will exclusively discuss on experiment and results of the utilization of developed adaptive contact model to compensate the time delay. Figure 8.10 describes the experiment scenario. The operator at the local station relied on the simulated visual feedback from the adaptive contact model to determine how the incision should be done by the teleoperator. In the remote environment, the teleoperator was controlled to perform an incision in the test object. The time delay of $300\,ms$ was given as the previous experiment.

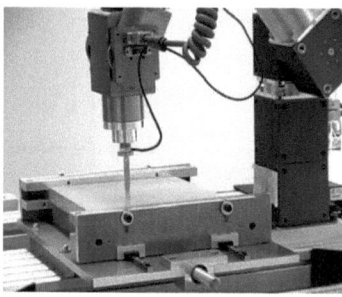

(a) Operator at local station   (b) Teleoperator in remote environment

Figure 8.10: Utilization of the adaptive contact model in an incision teleoperation with a large time delay

The adaptive contact model was modeled so that the virtual object replicates the actual test object in terms of geometry and material properties. The calculation of the dynamic geometry deformation employed the XFEM approach. The sampling rate of adaptive empirical incision force model was $500\,Hz$ which is adequate for a seamless force rendering. The measured incision force from the force sensor on the teleoperator end-effector was used coupling with the adaptive parameter identification algorithm.

Figure 8.11 demonstrates the $300\,ms$-delayed arrival of the actual incision force from the sensor as the previous experiment. In contrast, the delayed force signal was not forwarded to the operator. It was compensated by the adaptive contact model as soon as the incision process took place. As a result, the cut-depth did not exhibit an oscillation since the synchronization between hand's movement and the force perception was preserved. During the incision, the predicted incision force was validated with the delayed but correct force signal from the sensor which allowed a continuous correction to the empirical incision force model if an error between them existed.

For evaluation propose, the predicted incision force is plot against the delayed measurement signal in Figure 8.12. The function of the adaptive parameter identification algorithm can be seen.

8.2 Time delay compensation utilizing contact model

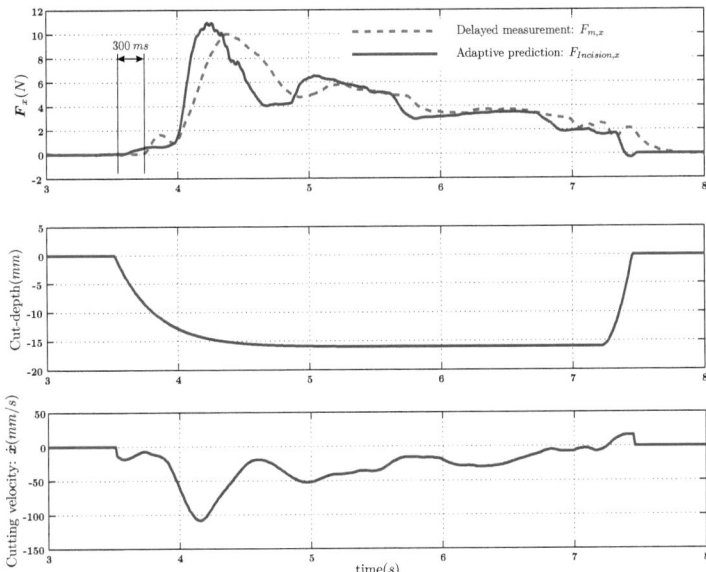

Figure 8.11: Adaptive incision force model provides an accurate and stable telepresence operation during an incision

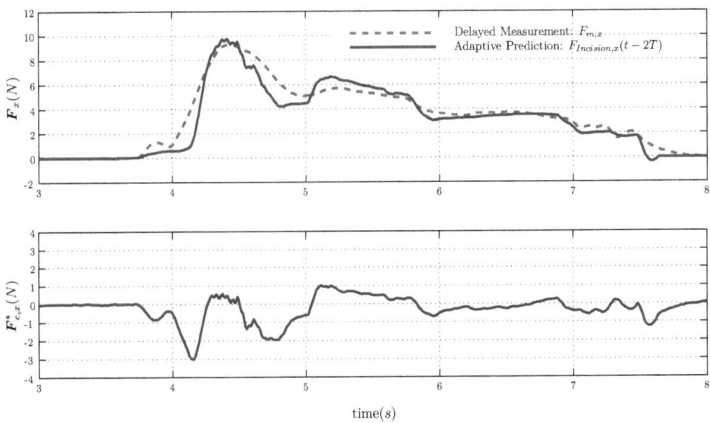

Figure 8.12: Accuracy improvement in incision force error $F^*_{e,x}$ with adaptive parameter identification algorithm

8.2 Time delay compensation utilizing contact model

The algorithm acknowledged 300 ms after the incision began that the incision force model was not correct with a large amount force error $F^*_{e,x}$. The algorithm responded by adjusting the coefficient τ_x in the empirical incision force model allowing $F^*_{e,x}$ to decline as the incision was carried out without an interruption. The proposed algorithm was successfully predicted an incision force without significant delayed. It provided a transparent correction during the actual incision was taken place. As a result, a stable and comparable realistic telepresence operation was obtained. Figure 8.13 illustrates the cut in the test object which does not exhibit the damage due to existing large time delay compared to Figure 8.9. On the other hand, a similar result found in 8.7 by the reference experiment was instead achieved.

Figure 8.13: Adaptive contact model maintains the stability with a result similar to figure 8.7

Although it is not a focus of this work to interpret the success rate of the teleoperation with a psychological assessment, it is necessary to discuss that the results presented in this chapter subject to the human-factor of the operators. Therefore, the results from each experiment should be relatively evaluated with the results from that particular operator and not from the others.

In the second experiment, where the 300 ms was given, the better results were observed from operators with an understanding of telepresence and time delay. This group of operators provided a certain level of extra force reducing the eradicate movement of the haptic device. Therefore, the delayed force feedback in this case were neglected as a perception but treated as an interference instead. Since it was not constrained how the operators should make an incision, some operators moved their hands slowly enough to obtain a reasonable level of synchronization even though the force signal was delayed severely. The time delay was therefore not a significant issue in this case. These individual abilities of handling a machine control interface depend on personal experiences which result in different levels of success and quality of the achievements.

8.2 Time delay compensation utilizing contact model

It is arguable that some telepresence operations may be successful carried out without a force perception. Specifically, when the force is distorted, such as from time delay. However, this may not be correct for a medical teleoperation. The demand of an accurate force feedback especially for a surgical teleoperation was studied by WAGNER ET AL. [97]. It is discussed that the force feedback does not only improve the rate of success significantly in a medical task handling but more importantly prevent a damage to the scenery. MORRIS ET AL. presented an evidence of benefits of a task training gained from a combination of visual and haptic [59]. The work finds that a visuohaptic training is significantly more accurate than a visual or haptic training alone.

From discussions during the experiments in this work, it can be concluded with confidence that a correct, or quantitatively speaking, realistic telepresence experience is always preferred. An intuitive telepresence experience proves to help operators to adjust themselves faster to telepresence control interface while allowing their personal skills to be expressed naturally and with confident. The utilization of the proposed adaptive contact model fulfills its role by substantially provide the operators with an accurate incision force even the communication channel is subjected to a large time delay.

9 Conclusion and future works

9.1 Discussion

In this work, the time delay problem in a medical telepresence system specifically during an incision process is addressed. The compensation of time delay with an adaptive contact model is proposed. The adaptive contact model is developed to simulate a dynamic geometry deformation for the visualization of a soft body and the corresponding incision force in real-time. The adaptive parameter identification algorithm in the contact model provides an on-line verification capability of the simulated incision force during the actual incision in telepresence operation.

The development begins with a thorough analysis of the incision mechanics in terms of continuum mechanics in chapter 3. A discussion is made on the configurational changes of material points in a soft body when a cut evolves. It leads to the need of a remapping of the dynamic governing system of equation. To realize a dynamic geometry deformation simulation, a continuum soft body is discretized into finite numbers of smaller domains. FEM and XFEM are two approaches investigated in this work.

In chapter 4, in order to satisfy the remapping of configurational changes due to material point separation, FEM requires a reconsideration of the spatial discretization called remeshing. It results in an update to the Lagrange mesh at each simulation step. An example of remeshing is also developed allowing the cut to evolve on the edge of tetrahedron elements. FEM with remeshing can be seen to treat the material discontinuity from the cut in a global domain. Therefore, the governing system of equations is reassembled after the current topology before the remapped boundary conditions can be applied.

On the other hand, in chapter 5, XFEM approach uses a shifted function to approximate a cut as a material discontinuity to an element. The rest of the elements in the initial Lagrange mesh remains unchanged. This approach enables the governing system of equation to expand systematically without the initial governing system of equations to be completely rearranged.

In chapter 6, the discussion on the practicality of FEM and XFEM as a development tool for a incision contact model is brought. In contrary to FEM, XFEM demonstrates a good flexibility in handling a dynamic simulation problem with discontinuities. Therefore, XFEM is evaluated to be more practical and suitable for a contact model for incision simulation.

The incision force from the scalpel is a nodal contact force acting on the element which is being cut. The governing system of equations require the incision force to be given to fulfill the Neumann boundary conditions. Therefore, in chapter 7, an empirical incision force is developed after the friction model. Considering an actual surgical telepresence scenery, a perfect modeling and parameterization of the proposed incision force model can be difficult to achieve. The optimized adaptive parameter identification algorithm is proposed in this work for a constant verification of the accuracy of the contact model during the incision in an actual telepresence operation. The algorithm proves a substantial improvement in the accuracy while transparently provides a correct incision force perception to the human operators.

In contrast to general telepresence systems, the experimental telepresence system is developed to subject to a large time delay ($300\,ms$) in communication in chapter 8. This setup purposefully provokes the instability during the actual incision. In an experiment, the instability is observed from a non-synchronization of the hand's movement and the force feedback. The operators are subjected to perceive this false contact force while penetrates the test object. The experiment shows that instead the force feedback provides the human operators a comprehend contact information, it interferences the process. With a scalpel at the end-effector of the teleoperator, the damage occurred to the test object is inevitable.

In another experiment, the time delay is completely compensated from the control route of the telepresence system using the predicted visualization and force signal obtained from the adaptive contact model developed in this work. Thus, a stable incision with a telepresence system subjected to a large time delay is achieved.

9.2 Outlooks

The proposed contact model for incision process developed in this work can be further developed to be an interactive surgical simulator framework similar to the SOFA framework [10]. From the derivations seen in this work, XFEM is a convincing approach for the modeling of a contact mechanics especially with the different types of discontinuities at the same time. The contact library shall be invented to support an implementation of a contact model with a comprehensive archive of model and biomechanical properties of different organs can be obtained similar to the work of AGACHE ET AL. [9], SCHILLHUBER [71] and ZHAO [100].

Three-dimensional geometry of patient's organs can nowadays easily obtained from conventional Computerized Tomography (CT), Magnetic Resonance Imaging (MRI) techniques. It can be combined with the mechanical properties supplied by the contact library to develop an approximate contact model of a patient. This contact model with realistic visual graphic and haptic interaction can provide an intensified training of medical residents and doctors which induces failure prevention and a more efficient surgical operation.

9.2 Outlooks

Telepresence technology is a widely embraced application. In medical fields, the telesurgery has been proving and slowly accepted for its potential as a better option to the patient. However, before this goal can be accomplished, some technical limitations such as a successful skill transfer between a doctor and a robotic system must be realized. This can depend on a flawless feedback perception provided by the robotic system to the surgeon. The adaptive contact model developed in this work proved to provide an accurate incision force. However, this can be achieved because the experiment scenery such as the position of the test object does not change. In an actual surgery, the contact model can encounter a difficulty of the treatment of the boundary conditions. This problem can be solved by using a camera system to calculate the positions of the object of interest while constantly update the Dirichlet boundary conditions in the simulation.

The experiments in this work did not involve in an psychological assessment. The results conclude that the accurate incision force is preferred. Nevertheless, since the human factor or preference to the interaction with the robot control interface can be varied, a systematic evaluation considering human preference with the rate of success can provide a new information for the improvement of the contact model and the control strategy of the telepresence system.

Within Collaborative Research Centre 453, it has been discussed of the integration of the various methods to provide a stable telepresence system in different purposes. The combination of the proposed adaptive contact model with the transparent data reduction [43] and data compression in communication [88] is a logical approach. The contact model can supplement the simulation results in cases the data is completely lost. Therefore, the teleoperation can be carried out without an interruption.

A Mechanical properties of silicone

Table A.1: Table of properties: Silicone SI60GB. Source: Breddermann Kunstharze

Mechanical properties	Value	Unit	Test method
Density	1.05	g/cm^2	calculated
Ultimate tensile strength	60×10^5	N/m^2	DIN 53 504
Ultimate elongation	330	%	DIN 53 504
Shrinkage	< 0.1	%	7d
Color	translucent, colorless		

Table A.2: Table of properties: Silicone SI6GB. Source: Breddermann Kunstharze

Mechanical properties	Value	Unit	Test method
Density	1.3	g/cm^2	calculated
Young's modulus	3.85×10^5	N/m^2	See section 8.1.3
Ultimate tensile strength	23.6×10^5	N/m^2	DIN 53 504
Ultimate elongation	292	%	DIN 53 504
Shrinkage	< 0.2	%	7d
Color	light green		

Bibliography

[1] *Sonderforschungsbereich 453 (SFB 453)*. Website. http://www.lrz.de/~t8241ad/webserver/webdata/index.html

[2] *Khronos Group - Open Standard for Parallel Programming of Heterogeneous Systems : OpenCL*. Website. www.khronos.org/opencl/. Version: 2011

[3] *Nvidia Technology - Compute Unified Device Architecture : CUDA*. Website. www.nvidia.com. Version: 2011

[4] *Open Graphics Library: OpenGL*. Website. www.opengl.org. Version: 2011

[5] *Open Multi-Processing : OpenMP*. Website. www.openmp.org. Version: 2011

[6] *SCHUNK GmbH & Co. KG*. Website. www.opengl.org. Version: 2011

[7] *Sensable Technologies*. Website. www.sensable.com. Version: 2011

[8] ABE, N. ; YAMANAKA, K.: Smith predictor control and internal model control - a tutorial. In: *SICE 2003 Annual Conference* vol. 2, 2003, pp. 1383–1387

[9] AGACHE, P.G. ; MONNEUR, C. ; LEVEQUE, J. L. ; RIGAL, J. D.: Mechanical Properties and Young's Modulus of Human Skin in Vivo. In: *Archives of Dermatological Research* 269 (1980), pp. 221–232

[10] ALLARD, J. ; COTIN, S. ; FAURE, F. ; BENSOUSSAN, P. ; POYER, F. ; DURIEZ, C. ; DELINGETTE, H. ; JEŘÁBKOVÁ, L. G.: SOFA - an Open Source Framework for Medical Simulation. In: *Medecine Meets Virtual Reality* (2007), pp. 13–18

[11] ARMSTRONG-HÉLOUVRY, B. ; DUPONT, P. ; WIT, C. C.: A Survey of Models, Analysis Tools and Compensation Methods for the Control of Machines with Friction. In: *Automatica* 30 (1994), pp. 1083–1138

[12] ASHRAFIUON, H.: Design Optimization of Aircraft Engine-Mount Systems. In: *Vibration and Control of Mechanical Systems* 61 (1993), pp. 117–131

[13] ch. Efficient algorithms for layer assignment problems. In: BATHE, K.-J: *Finite-Elemente-Methoden*. San Francisco, CA : Freeman, 1973, pp. 63–83

[14] BELYTSCHKO, T. ; BLACK, T.: Elastic Crack Growth in Finite Elements With Minimal Remeshing. In: *International Journal for Numerical Methods in Engineering* 45 (1999), pp. 601–620

[15] BELYTSCHKO, T. ; LIU, W. K. ; MORAN, B.: *Nonlinear Finite Elements for Continua and Stuctures*. John Wiley & Sons, Inc., 2000

[16] BORDAS, Ss ; LEGAY, A. ; GRAVOUIL, A.: *X-FEM Mini-Course*. 2007. – Lecture note from EPSFL short-course

Bibliography

[17] BREMER, H. ; PFEIFFER, F.: *Dynamik und Regelung mechanischer Systeme.* Teuber Verlag, 1988

[18] BREMER, H. ; PFEIFFER, F.: *Elastische Mehrkörpersysteme.* Teuber Verlag, 1992

[19] BRUYNS, C. D. ; MONTGOMERY, K.: Generalized Interactiions Using Virtual Tools within the Spring Framework: Cutting. In: *Medicine Meets Virtual Reality* 02/10 (2002), pp. 79–85

[20] BURNETT, D. S.: *Finite Element Analysis.* Addison-Wesley, 1987

[21] CHESSA, J. ; BELYTSCHKO, T.: An enriched finite element method and level sets for axisymmetric two-phase flow with surface tension. In: *International Journal for Numerical Methods in Engineering* 58 (2003), pp. 2041–2064

[22] CLARKE, S. ; SCHILLHUBER, G. ; ZAEH, M. F. ; ULBRICH, H.: The Effect of Simulated Inertia and Force Prediction on Delayed Telepresence. In: *Presence: Teleoperators and Virtual Environments* 16 (2007), October, nr. 5, pp. 543-558. – Special Topic: Teleoperators and Virtual Environments

[23] CLOUGH, R. ; PENZIEN, J. ; EDITION 2nd (ed.): *Dynamics of Structures.* MeGraw-Hill, 1993

[24] CONTI, F. ; BARBAGLI, F. ; MORRIS, D. ; SEWELL, C.: CHAI 3D - An Open-Source Library for the Rapid Development of Haptic Scenes. In: *IEEE World Haptics*, 2005

[25] DOLBOW, J.: *An Extended Finite Element Method with Discontinuous Enrichment for Applied Mechanics*, Northwestern University, Evanston, IL, Diss., 1999

[26] EBERHARD, P.: *Kontaktuntersuchungen durch hybride Mehrkörpersystem / Finite Elemente Simulationen.* Shaker Verlag, 2000

[27] ENGELKE, R.: *Modellierung und Optimierung von Robotern mit einseitigen Bindungen und lokalen Verspannungen.* München, Germany, Technische Universität München, Fakultät für Maschinenewesen, Diss., 2008

[28] FITE, K.B. ; GOLDFARB, M. ; RUBIO, A.: Transparent telemanipulation in the presence of time delay. In: *Advanced Intelligent Mechatronics, 2003. AIM 2003. Proceedings. 2003 IEEE/ASME International Conference on* vol. 1, 2003, pp. 254 – 259 vol.1

[29] FRIES, T. P. ; BELYTSCHKO, T.: The Instrinsic XFEM: A Method for Arbitrary Discontinuities Without Additional Unknowns. In: *International Journal for Numerical Methods in Engineering* 68 (2006), pp. 1358–1385

[30] FULLER, D. D.: Theory and practice of lubrication for engineers. In: *Journal of Synthetic Lubrication* 1, pp. 314

[31] GOANGSEUP, Z. ; BELYTSCHKO, T.: New Crack-Tip Elements for XFEM and Applications to Cohesive Cracks. In: *International Journal for Numerical Methods in Engineering* 57 (2003), pp. 2221–2240

[32] GOLLE, A. ; ULBRICH, H.: Contact Models for Real-Time-Simulation in Telepresence Applications. In: *Inproceeding of EuroHaptics 2004*, 2004, pp. 381–384

[33] GOLLE, A. ; ULBRICH, H. ; PFEIFFER, F.: Real-time simulation of non-smooth contacts in telepresence. In: *Industrial Technology, 2003 IEEE International Conference on*, 2003 (2), pp. 675–679

[34] GREEN, P. S. ; HILL, J. W. ; JENSEN, J. F. ; SHAH, A.: Telepresence Surgery. In: *IEEE Engineering in Medicine and Biology* 14 (1995), pp. 324–329

[35] GROSS, S. ; REUSKEN, A.: An extended pressure finite element space for two-phase incompressible flows with surface tension. In: *Journal of Computational Physics* 224 (2007), pp. 40–58

[36] GÜRSES, E. ; MIEHE, C.: A Computational Framework of Three-Dimensional Configurational-Force-Driven Brittle Crack Propagation. In: *Computer methods in applied mechanics and engineering* (2008)

[37] HAHN, C.: *Models, Algorithms and Software Concepts for Contact and Fragmentation in Computational Solid Mechanics*, Faculty of Civil Engineering, Universität Hannover, Diss., 2005

[38] HAHN, J. ; EDISON, T. ; EDGAR, T. F.: A Note on Stability Analysis Using Bode Plots. In: *ChE classroom* (2001)

[39] HANSBO, A. ; HANSBO, P.: A Finite Element Method for the Simulation of Strong and Weak Discontinuities in Solid Mechanics. In: *Computer methods in applied mechanics and engineering* 193 (2004), pp. 3523 – 3540

[40] HEIMANN, B. ; GERTH, W. ; POPP, K.: *Mechatronik: Komponent - Methoden - Beispiele*. 2. Fachbuchverlag Leipzig, 2000

[41] ch. 14. In: HELD, R. ; DURLACH, N.l: *TELEPRESENCE, TIME DELAY, AND ADAPTATION*. Taylor and Francis, 1991, pp. 232–246

[42] HIRCHE, S. ; BAUER, A. ; BUSS, M.: Transparency of Haptic Telepresence Systems with Constant Time Delay. In: *IEEE Conference on Control Applications*, 2005

[43] HIRSCHE, S. ; BUSS, M.: Transparent Data Reduction in Networked Telepresence and Teleaction Systems Part II: Time-Delayed Communication. In: *Presence: Teleoperators and Virtual Environments* 16, No. 5 (2007), pp. 532–542

[44] HIRZINGER, G. ; HAGN, U.: Flexible Heart Surgery. In: *German Research (Magazine of the German Research Foundation DFG)* 1 (2010). – Available online

[45] HRENIKOFF, A.: Solution of probelms in eleasticity by the framwork method. In: *ASME Journal of Applied Mechanics* A8 (1941), pp. 169–175

[46] JEŘÁBKOVÁ, L.: *Interactive Cutting of Finite Elements based Deformable Objects in Virtual Environment*, Rheinisch-Westfälischen Technischen Hochschule Aachen, Faculty of Mathematics, Computer Science and Natural Sciences, Dissertation, 2007

Bibliography

[47] JOHNSTON, M. D. ; RABE, K. J.: Integrated Planning for Telepresence with Time Delays. In: *2nd IEEE International Conference on Space Mission Challenges for Information Technology SMCIT06*, 2006, pp. 140–146

[48] KNOLL, A.: Toward High-Definition Telepresence. In: *Presence: Teleoperators and Virtual Environments* 16 (2007), nr. 5

[49] KUNDU, K. ; OLANO, Ol: Tissue resection using delayed updates in a tetrahedral mesh. In: *Studies in Health Technology and Informatics* (2007)

[50] LANDZETTEL, K. ; PREUSCHE, C. ; ALBU-SCHAFFER ; A. REINTSEMA, D. ; REBELE, D. ; HIRZINGER, G.: Robotic On-Orbit Servicing - DLR's Experience and Perspective. In: *IEEE/RSJ International Conference on Intelligent Robots and Systems*, 2006, pp. 4587

[51] LANG, C. B. ; PUCKER, N.: *Mathematische Methoden in der Physik*. 2. Spektrum Akademischer Verlag

[52] LAURSEN, T. A. ; SIMO, J. C.: A Continuum-Based Finite Element Formulation for the Implicit Solution of Multibody, Large Deformation-Frictional Contact Problems. In: *International Journal for Numerical Methods in Engineering* 36 (2005), pp. 3451–3485

[53] MAYER, H.: *Human Machine Skill Transfer in Robot Assisted, Minimally Invasive Surgery*. Munich, Germany, Technische Universität München, Diss., 2008

[54] MCHENRY, D.: A lattice analogy for the solution of plane stress problems. In: *International Journal of Civil Engineering* 21 (1943), pp. 59–82

[55] MITSUISHI, M. ; IIZUKA, Y. ; WATANABE, H. ; HASHIZUME, H. ; FUJIWARA, K.: Remote operation of a micro-surgical system. In: *IEEE International Conference on Robotics and Automation* vol. 2, 1998, pp. 1013–1019

[56] MOËS, N. ; DOLBOW, J. ; BELYTSCHKO, T.: A Finite Element Method For Crack Growth Without Remeshing. In: *International Journal for Numerical Methods in Engineering* 46 (1999), pp. 131–150

[57] MOËS, N. ; SUKUMAR, N. ; MORAN, B. ; BELYTSCHKO, T.: An Extended Finite Element Method (X-FEM) For Two- and Threes-Dimensional Crack Modeling. In: *European Congress on Computational Methods in Applied Sciences and Engineering ECCOMAS 2000*, 2000

[58] MOHAMMADI, S.: *Extended finite element method for fracture analysis of structures*. Blackwell Publishing Ltd, 2008

[59] MORRIS, D. ; TAN, H. ; BARBAGLI, F. ; CHANG, T. ; SALISBURY, K.: Haptic Feedback Enhances Force Skill LearningT̂, to appear. In: *Proceedings of the 2007 World Haptics Conference*, 2007, pp. 22–24

[60] NEWMARK, N. M.: Numerical methods of analysis in bars, plates and elastic bodies. In: *Numerical Methods in Analysis in Engineering* (1949)

[61] NEWMARK, N. M.: A Method of Computation for Structural Dynamics. In: *Journal of Engineering Mechanics Division* 85 (1959), pp. 67–94

Bibliography

[62] O'HARA, G. P.: Mechanical Properties of Silicone Rubber in a Closed Volume. In: *US ARMY ARMAMENT RESEARCH AND.DEVELOPMENT CENTER* (1984)

[63] ONO, K. ; SCHILLHUBER, G. ; ULBRICH, H.: XFEM Approach to Real-Time Incision Haptic Feedback for Surgical Simulation. In: *International Conference on Extended Finite Element Method - Recent Developments and Applications*, 2009

[64] ORTMAIER, T. ; WEISS, H. ; HIRZINGER, G.: Telepresence and Teleaction in Minimal Invasive Surgery. In: *Proceedings of the Robotik VDI-Berichte*, 2004

[65] PEER, A. ; PONGRAC, H. ; BUSS, M.: Influence of Varied Human Movement Control on Task Performance and Feeling of Telepresence. In: *Presence: Teleoperators and Virtual Environments* 19 (2010), pp. 463–481

[66] PFEIFFER, F. ; GOLOCKER, C.: *Multibody Dynamics with Unilateral Contacts*. John Wiley & Sons, Inc., 1996

[67] PFEIFFER, F. ; REITHMEIER, E.: *Roboerdynamik*. Teubner Verlag, 1987

[68] PRESS, W. H. ; TEUKOLSKY, S. A. ; VETTERLING, W. T. ; FLANNERY, B. P.: *Numerical Recipes in C : The Art of Scientific Computing, Second Edition*. CAMBRIDGE UNIVERSITY PRESS, 1997

[69] PREUSCHE, C. ; REINTSEMA, D. ; LANDZETTEL, K. ; HIRZINGER, G.: Robotics Component Verification on ISS ROKVISS - Preliminary Results for Telepresence. In: *IEEE/RSJ International Conference on Intelligent Robots and Systems*, 2006, pp. 4595

[70] REINHART, G. ; CLARKE, S. ; PETZOLD, B. ; SCHILP, J.: Telepresence as a Solution to Manual Micro-Assembly. In: *CIRP Annals* 53 (2004), pp. 21–24

[71] SCHILLHUBER, G.: *Robotergestützte Modellidentifikation und Simulation von deformierbaren Körpern für haptische Anwendungen*. Munich, Germany, Technische Universität München, Faculty of Mechanical Engineering, Diss., 2008

[72] SCHILLHUBER, G. ; ULBRICH, H.: Real-time FEM for haptic applications under consideration of human perception. In: *the 78th Annual Meeting of the Gesellschaft für Angewandte Mathematik und Mechanik e.V.* Zurich, Switzerland, 2007

[73] SCHILLHUBER, G. ; ULBRICH, H.: Haptic Simulation of Deformable Bodies with Consideration of the Human Sensation of Continuous Forces. In: *The 16th Symposium on Haptic Interfaces for Virtual Environments and Teleoperator Systems*. Reno, Nevada, USA, 2008

[74] SELA, G. ; SUBAG, J. ; LINDBLAD, A. ; ALBOCHER, D. ; SCHEIN, S. ; ELBER, G.: Real-time haptic incision simulation using FEM-based discontinuous free form deformation. In: *Proceedings of the 2006 ACM symposium on Solid and physical modeling*. New York, NY, USA : ACM, 2006 (SPM '06). – ISBN 1-59593-358-1, 75–84

Bibliography

[75] SEWELL, C.: *Automatic Performance Evaluation in Surgical Simulation: Providing Metrics and Constuctive Feedback in a Haptic Mastoidectormy Simulator.* VDM Verlag, 2008

[76] SHERIDAN, T. B.: Telerobotics, automation, and human supervisory control. In: *MIT Press* (1992)

[77] SHIBAHARA, M. ; SERIZAWA, H. ; MURAKAWA, H.: Finite Element Method for Hot Cracking Using Temperature Dependent Interface Element (ReportII) - Mechanical Study of Houldcroft Test. In: *Transaction of Joining and Welding Research Institute, Osaka University* 1 (2000), pp. 59–64

[78] SHIBAHARA, M. ; SERIZAWA, H. ; MURAKAWA, H.: Finite Element Analysis of Hot Cracking Under Welding Using Temperature-Dependent Interface Element. In: *In Proceedings of the 11th International Offshore and Polar Engineering Conference*, 2001

[79] SI, H. ; GÄRTNER, K.: An Algorithm for Three-Dimensional Constrained Delaunay Tetrahedralizations. In: *Proceeding of the Fourth International Conference on Engineering Computational Technology*, 2004

[80] SI, H. ; GÄRTNER, K.: Meshing Piecewise Linear Complexes by Constrained Delaunay Tetrahedralizations. In: *Proceeding of the 14th International Meshing Roundtable*, 2005

[81] SICILIANO, B. ; SCIAVICCO, L. ; VILLANI, L. ; ORIOLO, G.: *Robotics Modelling, Planning and Control.* Springer, 2008

[82] SLOTINE, J.-J. E. ; LI, W.: Adaptive manipulator control: A case study. In: *Automatic Control, IEEE Transactions on* 33 (1988), November, nr. 11, pp. 995–1003. – ISSN 0018-9286

[83] SLOTINE, J.-J. E. ; LI, W.: Composite adaptive control of robot manipulators. In: *Automatica* 25 (1989), July, pp. 509–519. – ISSN 0005-1098

[84] SMITH, A. C. ; HASHTRUDI-ZAAD, K.: Smith Predictor Type Control Architectures for Time Delayed Teleoperation. In: *International Journal of Robotics Research* 25 (2006), August, nr. 8, pp. 797–818

[85] SMITH, O. J. M.: A controller to overcome dead-time. In: *ISA Transactions* 6 (1959), pp. 28–33

[86] SOUTHWELL, R. V. ; 1 (ed.): *Relaxation Methods in Theorectical Physics.* Clarendon Press, 1946

[87] STAUB, C. ; KNOLL, A. ; OSA, T. ; BAUERNSCHMITT., R.: Autonomous high precision positioning of surgical instruments in robot-assisted minimally invasive surgery under visual guidance. In: *IEEE International Conference on Autonomic and Autonomous Systems.* Cancun, Mexico, March 2010, pp. 64–69

[88] STEINBACH, E. ; HIRCHE, S. ; KAMMERL, J. ; VITTORIAS, I. ; CHAUDHARI, R. G.: Haptic Data Compression and Communication for Telepresence and Teleaction. In: *IEEE Signal Processing Magazine* 28 (2011), January, nr. 1, pp. 87–96

Bibliography

[89] SUKUMAR, N. ; CHOPP, D.L. ; MOËS, N. ; BELYTSCHKO, T.: Modeling Hoels and Inclusions by Level Sets in the Extended Finite-Element Method. In: *Computer Methods in Applied Mechanics and Engineering* 190 (2000), pp. 6183–6200

[90] TAVAKOLI, M. ; HOWE, Robert D.: Haptic Effects of Surgical Teleoperator Flexibility. In: *Int. J. Rob. Res.* 28 (2009), October, pp. 1289–1302. – ISSN 0278-3649

[91] TERZOPOULOS, D.i ; PLATT, J. ; BARR, A. ; FLEISCHER, K.: Elastically deformable models. In: *14th International Conference on Computer Graphics and Interactive Techniques*, 1987 (4 21), pp. 205–214

[92] THOMPSON, T. J. ; BAUMGARTEN, J. R.: A Brief Study of Passive Viscous Damping for the Spice Bulkhead Structure. In: *Vibration and Control of Mechanical Systems* 61 (1993), pp. 133–145

[93] TRÄDEGÅRD, A. ; NILSSON, F. ; ÖSTLUND, S.: FEM-Remeshing Technique Applied to Crack Growth Problems. In: *Computner Methods in Applied Mechanics and Engineering* 160 (1998), pp. 115–131

[94] TURNER, M. J. ; CLOUGH, R. W. ; MARTIN, H. C. ; TOPP, L. J.: Stiffness and deflection analysis of complex structures. In: *Aeronautic Science* 23 (1956), pp. 805–823

[95] ULBRICH, H.: *Maschinendynamik*. Teuber Verlag, 1996

[96] VIGNERON, L. M. ; VERLY, J. G. ; WARFIELD, S. K.: On Extended Finite Element Method (XFEM) for Modelling of Organ Deformations Associated with Surgical Cuts. In: *Lecture Notes in Computer Science: Medical Simulation* 3078 (2004), pp. 134–143

[97] WAGNER, C. R. ; STYLOPOULOS, N. ; HOWE, R. D.: The Role of Force Feedback in Surgery: Analysis of Blunt Dissection. In: *the 10th Symposium on Haptic Interfaces for Virtual Environment and Teleoperator Systems*, 2002

[98] WRIGGERS, P.: *Nichtlineare Finite-Element-Methoden*. Springer Verlag, 2001

[99] WRIGGERS, P.: *Computaional Contact Mechanics*. John Wiley & Sons, 2002

[100] ZHAO, C.: *Real Time Haptic Simulation of Deformable Bodies*. Munich, Germany, Technische Universität München, Faculty of Mechanical Engineering, Diss., 2010

[101] ZIENKIEWICZ, O.C. ; TAYLOR, R.L.: *The Finite Element Method for Solid and Structural Mechanics*. Elsevier Butterworth Heinemann, 2005

[102] ZIENKIEWICZ, O.C. ; TAYLOR, R.L. ; ZHU, J.Z.: *The Finite Element Method: Its Basis and Fundamentals*. Elsevier Butterworth Heinemann, 2005

i want morebooks!

Buy your books fast and straightforward online - at one of world's fastest growing online book stores! Environmentally sound due to Print-on-Demand technologies.

Buy your books online at
www.get-morebooks.com

Kaufen Sie Ihre Bücher schnell und unkompliziert online – auf einer der am schnellsten wachsenden Buchhandelsplattformen weltweit! Dank Print-On-Demand umwelt- und ressourcenschonend produziert.

Bücher schneller online kaufen
www.morebooks.de

 VDM Verlagsservicegesellschaft mbH
Heinrich-Böcking-Str. 6-8   Telefon: +49 681 3720 174   info@vdm-vsg.de
D - 66121 Saarbrücken   Telefax: +49 681 3720 1749 www.vdm-vsg.de

Printed by Books on Demand GmbH, Norderstedt / Germany